Frederic P. Miller, Agnes
John McB

Environmental Ethics

Environmental philosophy, Environmental
law, Environmental sociology, Ecotheology,
Ecological economics, Ecology,
Environmental geography, Clearcutting,
Internal combustion engine

Alphascript Publishing

Imprint

Permission is granted to copy, distribute and/or modify this document under the terms of the GNU Free Documentation License, Version 1.2 or any later version published by the Free Software Foundation; with no Invariant Sections, with the Front-Cover Texts, and with the Back-Cover Texts. A copy of the license is included in the section entitled "GNU Free Documentation License".

All parts of this book are extracted from Wikipedia, the free encyclopedia (www.wikipedia.org).

You can get detailed informations about the authors of this collection of articles at the end of this book. The editors (Ed.) of this book are no authors. They have not modified or extended the original texts.

Pictures published in this book can be under different licences than the GNU Free Documentation License. You can get detailed informations about the authors and licences of pictures at the end of this book.

The content of this book was generated collaboratively by volunteers. Please be advised that nothing found here has necessarily been reviewed by people with the expertise required to provide you with complete, accurate or reliable information. Some information in this book maybe misleading or wrong. The Publisher does not guarantee the validity of the information found here. If you need specific advice (f.e. in fields of medical, legal, financial, or risk management questions) please contact a professional who is licensed or knowledgeable in that area.

Any brand names and product names mentioned in this book are subject to trademark, brand or patent protection and are trademarks or registered trademarks of their respective holders. The use of brand names, product names, common names, trade names, product descriptions etc. even without a particular marking in this works is in no way to be construed to mean that such names may be regarded as unrestricted in respect of trademark and brand protection legislation and could thus be used by anyone.

Cover image: www.ingimage.com
Concerning the licence of the cover image please contact ingimage.

Publisher:
Alphascript Publishing is a trademark of
VDM Publishing House Ltd.,17 Rue Meldrum, Beau Bassin,1713-01 Mauritius
Email: info@vdm-publishing-house.com
Website: www.vdm-publishing-house.com

Published in 2010

Printed in: U.S.A., U.K., Germany. This book was not produced in Mauritius.

ISBN: 978-613-0-66933-1

Contents

Articles

References

Environmental ethics

Environmental ethics is the part of → environmental philosophy which considers extending the traditional boundaries of ethics from solely including humans to including the non-human world. It exerts influence on a large range of disciplines including → law, → sociology, → theology, → economics, → ecology and → geography.

There are many ethical decisions that human beings make with respect to the environment. For example:

- Should we continue to clear cut forests for the sake of human consumption?
- Should we continue to propagate?
- Should we continue to make → gasoline powered vehicles?
- What environmental obligations do we need to keep for future generations?
- Is it right for humans to knowingly cause the → extinction of a species for the convenience of humanity?

The academic field of environmental ethics grew up in response to the work of scientists such as → Rachel Carson and events such as the first → Earth Day in 1970, when environmentalists started urging philosophers to consider the philosophical aspects of environmental problems. Two papers published in Science had a crucial impact: Lynn White's "The Historical Roots of our Ecologic Crisis"The_Historical_Roots_of_Our_Ecologic_Crisis The Historical Roots of our Ecologic Crisis]and [[Garrett HardiAlso influential was Garett Hardin's later essay called "Exploring New Ethics for Survival", as well as an essay by → Aldo Leopold in his → *A Sand County Almanac*, called "The Land Ethic," in which Leopold explicitly claimed that the roots of the ecological crisis were philosophical

The first international academic journals in this field emerged from North America in the late 1970s and early 1980s – the US-based journal, *Environmental Ethics* in 1979 and the Canadian based journal *The Trumpeter: Journal of Ecosophy* in 1983. The first British based journal of this kind, → *Environmental Values*, was launched in 1992.

Marshall's categories of environmental ethics

There have been a number of scholars who've tried to categorise the various ways the natural environment is valued. Alan Marshall and Michael Smith are two recent examples of this, as cited by Peter Vardy in "The Puzzle of Ethics".[1] For Marshall, three general ethical approaches have emerged over the last 40 years. Marshall uses the following terms to describe them: Libertarian Extension, the Ecologic Extension and Conservation Ethics.

(For more on Marshall's environmental ethics, see also: A. Marshall, 2002, The Unity of Nature, Imperial College Press: London. Alan Marshall is currently at the Univerzita Pavol Jozef Safarik, Slovakia).

Libertarian extension

Marshall's Libertarian extension echoes a civil liberty approach (i.e. a commitment to extend equal rights to all members of a community). In environmentalism, though, the community is generally thought to consist of non-humans as well as humans.

Andrew Brennan was an advocate of ecologic humanism (eco-humanism), the argument that all ontological entities, animate and in-animate, can be given ethical worth purely on the basis that they exist. The work of Arne Næss and his collaborator Sessions also falls under the libertarian extension, although they preferred the term "→ deep ecology." Deep ecology is the argument for the intrinsic value or inherent worth of the environment – the view that it is valuable in itself. Their argument, incidentally, falls under both the libertarian extension and the ecologic extension.

Peter Singer's work can be categorized under Marshall's 'libertarian extension'. He reasoned that the "expanding circle of moral worth" should be redrawn to include the rights of non-human animals, and to not do so would be guilty of speciesism. Singer found it difficult to accept the argument from intrinsic worth of a-biotic or "non-sentient" (non-conscious) entities, and concluded in his first edition of "Practical Ethics" that they should not be included in the expanding circle of moral worth.[2] This approach is essentially then, bio-centric. However, in a later edition of "Practical Ethics" after the work of Naess and Sessions, Singer admits that, although unconvinced by deep ecology, the argument from intrinsic value of non-sentient entities is plausible, but at best problematic. We shall see later that Singer actually advocated a humanist ethic.

Ecologic extension

Alan Marshall's category of ecologic extension places emphasis not on human rights but on the recognition of the fundamental interdependence of all biological (and some abiological) entities and their essential diversity. Where as Libertarian Extension can be thought of as flowing from a political reflection of the natural world, Ecologic Extension is best thought of as a scientific reflection of the natural world. Ecological Extension is roughly the same classification of Smith's eco-holism, and it argues for the intrinsic value inherent in collective ecological entities like ecosystems or the global environment as a whole entity. Holmes Rolston, among others, has taken this approach.

This category includes James Lovelock's Gaia hypothesis; the theory that the planet earth alters its geo-physiological structure over time in order to ensure the continuation of an equilibrium of evolving organic and inorganic matter. The planet is characterized as a unified, holistic entity with ethical worth of which the human race is of no particular significance in the long run.

Conservation ethics

Marshall's category of 'conservation ethics' is an extension of use-value into the non-human biological world. It focuses only on the worth of the environment in terms of its utility or usefulness to humans. It contrasts the intrinsic value ideas of 'deep ecology', hence is often referred to as 'shallow ecology', and generally argues for the preservation of the environment on the basis that it has extrinsic value − instrumental to the welfare of human beings. Conservation is therefore a means to an end and purely concerned with mankind and intergenerational considerations. It could be argued that it is this ethic that formed the underlying arguments proposed by Governments at the Kyoto summit in 1997 and three agreements reached in Rio in 1992.

Humanist theories

Following the bio-centric and eco-holist theory distinctions, Michael Smith further classifies Humanist theories as those that require a set of criteria for moral status and ethical worth, such as sentience. This applies to the work of Peter Singer who advocated a hierarchy of value similar to the one devised by Aristotle which relies on the ability to reason. This was Singer's solution to the problem that arises when attempting to determine the interests of a non-sentient entity such as a garden weed.

Singer also advocated the preservation of "world heritage sites," unspoilt parts of the world that acquire a "scarcity value" as they diminish over time. Their preservation is a bequest for future generations as they have been inherited from our ancestors and should be passed down to future generations so they can have the opportunity to decide whether to enjoy unspoilt countryside or an entirely urban landscape. A good example of a world heritage site would be the tropical rainforest, a very specialist ecosystem or climatic climax vegetation that has taken centuries to evolve. Clearing the rainforest for farmland often fails due to soil conditions, and once disturbed, can take thousands of years to regenerate.

Anthropocentrism

Anthropocentrism simply places humans at the centre of the universe; the human race must always be its own primary concern. It has become customary in the Western tradition to consider only our species when considering the environmental ethics of a situation. Therefore, everything else in existence should be evaluated in terms of its utility for us, thus committing speciesism. All environmental studies should include an assessment of the intrinsic value of non-human beings. [3] In fact, based on this very assumption, a philosophical article has explored recently the possibility of humans' willing extinction as a gesture toward other beings. [4] The authors refer to the idea as a thought experiment that should not be understood as a call for action.

What Anthropocentric theories do not allow for is the fact that a system of ethics formulated from a human perspective may not be entirely accurate; humans are not necessarily the centre of reality. The philosopher Baruch Spinoza argued that we tend to assess things wrongly in terms of their usefulness to us. Spinoza reasoned that if we were to look at things objectively we would discover that everything in the universe has a unique value. Likewise, it is possible that a human-centred or anthropocentric/androcentric ethic is not an accurate depiction of reality, and there is a bigger picture that we may or may not be able to understand from a human perspective.

Peter Vardy distinguished between two types of anthropocentrism. A strong thesis anthropocentric ethic argues that humans are at the center of reality and it is right for them to be so. Weak anthropocentrism, however, argues that reality can only be interpreted from a human point of view, thus humans have to be at the centre of reality as they see it.

Another point of view has been developed by Bryan Norton, who has become one of the essential actors of environmental ethics through his launching of what has become one of its dominant trends: environmental pragmatism. Environmental pragmatism refuses to take a stance in the dispute between the defenders of anthropocentrist ethics and the supporters of nonanthropocentrist ethics. Instead, Norton prefers to distinguish between *strong anthropocentrism* and *weak-or extended-anthropocentrism* and develops the idea that only the latter is capable of not under-estimating the diversity of instrumental values that humans may derive from the natural world[5] .

Status of the field

Environmental ethics became a subject of sustained academic philosophic reflection in the 1970s. Throughout the 1980s it remained marginalized within the discipline of philosophy, attracting the attention of a fairly small group of thinkers spread across the world.

Only after 1990 did the field gain institutional recognition at programs such as Colorado State, the University of Montana, Bowling Green State, and the University of North Texas. In 1991, Schumacher College of Dartington, England, was founded and now provides an MSc in Holistic Science.

These programs began to offer a masters degree with a specialty in environmental ethics/philosophy. Beginning in 2005 the Dept of Philosophy and Religion Studies at the University of North Texas offered a PhD program with a concentration in environmental ethics/philosophy.

See also

- → Biocentric individualism
- → Biocentrism
- Bioethics
- Climate ethics
- Conservation ethic
- → Conservation movement
- Deep Ecology
- EcoQuest (a series of two educational games)
- → Ecocentrism
- Ecofeminism
- → Ecological economics

- → Environmental movement
- → Environmentalism
- → Environmental skepticism
- Human ecology
- → List of environmental philosophers
- Population control
- Solastalgia
- Sustainability
- Terraforming
- Trail ethics
- → Van Rensselaer Potter
- Resource depletion
- Crop art

External links

- Bioethics Literature Database [6]
- Thesaurus Ethics in the Life Sciences [7]
- EnviroLink Library: Environmental Ethics [8] - online resource for environmental ethics information
- EnviroLink Forum - Environmental Ethics Discussion/Debate [9]
- Sustainable and Ethical Architecture Architectural Firm [10]
- Stanford Encyclopedia of Philosophy [11]
- Center for Environmental Philosophy [12]
- UNT Dept of Philosophy [13]
- Creation Care Reading Room [14]: Extensive online resources for environment and faith (Tyndale Seminary)

References

[1] Vardy, Peter. *The Puzzle of Ethics*.
[2] Singer, Peter. *Practical Ethics* (1 ed.).
[3] Singer, Peter. " *Environmental Values*. The Oxford Book of Travel Stories. Ed. Ian Marsh. Melbourne, Australia: Longman Chesire, 1991. 12-16.
[4] Tarik Kochi & Noam Ordan, "An Argument for the Global Suicide of Humanity". Borderlands, 2008, Vol. 3, 1-21 (http://www.borderlands. net.au/vol7no3_2008/kochiordan_argument.pdf).
[5] Afeissa, H. S. (2008) "The Transformative value of Ecological Pragmatism. An Introduction to the Work of Bryan G. Norton". *S.A.P.I.EN.S.* **1** (1) (http://sapiens.revues.org/index88.html)
[6] http://www.drze.de/BELIT?la=en-
[7] http://www.drze.de/BELIT/thesaurus?la=en-
[8] http://www.envirolink.org/newsearch.html?searchfor=ethics&x=0&y=0
[9] http://www.envirolink.org/forum
[10] http://www.ethicalarchitecture.co.uk/
[11] http://plato.stanford.edu/entries/ethics-environmental/
[12] http://www.cep.unt.edu/
[13] http://www.phil.unt.edu/
[14] http://www.tyndale.ca/sem/mtsmodular/viewpage.php?pid=73

Environmental philosophy

Environmental philosophy is a branch of philosophy that is concerned with the natural environment and humans' place within it.[1] Environmental philosophy includes → environmental ethics, environmental aesthetics, ecofeminism & environmental theology.[2] Some of the main areas of interest for environmental philosophers are:

- Defining environment and nature
- How to value the environment
- Moral status of animals and plants
- Endangered species
- Environmentalism and Deep Ecology
- Aesthetic value of nature
- Restoration of nature
- Consideration of future generations [1]

See also

- → List of environmental philosophers
- Natural philosophy

Further reading

- Armstrong, Susan, Richard Botzler. *Environmental Ethics: Divergence and Convergence,* McGraw-Hill, Inc., New York, New York.
- DesJardins, Joseph R., *Environmental Ethics* Wadsworth Publishing Company, ITP, An International Thomson Publishing Company, Belmont, California. A Division of Wadsworth, Inc.
- Derr, Patrick, G, Edward McNamara, 2003. *Case Studies in Environmental Ethics,* Bowman & Littlefield Publishers, Inc, Lanham, Maryland 20706 isbn = 0742531368
- Devall, W. and G. Sessions. 1985. *Deep Ecology: Living As if Nature Mattered,* Salt Lake City: Gibbs M. Smith, Inc.
- Foltz, Bruce V., Robert Frodeman. 2004. *Rethinking Nature,* Indiana University Press, 601 North Morton Street, Bloomington, IN 47404-3797 isbn = 0253217024
- Mannison, D., M. McRobbie, and R. Routley (ed), 1980. *Environmental Philosophy,* Australian National University
- Næss, A. 1989. *Ecology, Community and Lifestyle: Outline of an Ecosophy,* Translated by D. Rothenberg. Cambridge: Cambridge University Press.
- Pojman, Louis P., Paul Pojman. *Environmental Ethics,* Thomson-Wadsworth, United States
- Sherer, D., ed, Thomas Attig. 1983. *Ethics and the Environment,* Prentice-Hall, Inc., Englewood Cliffs, New Jersey 07632. isbn = 0132901633
- VanDeVeer, Donald, Christine Pierce. *The Environmental Ethics and Policy Book,* Wadsworth Publishing Company. An International Thomson Publishing Company
- Zimmerman, Michael E., J. Baird Callicott, George Sessions, Karen J. Warren, John Clark. 1993. *Environmental Philosophy: From Animal Rights to Radical Ecology,* Prentice-Hall, Inc., Englewood Cliffs, New Jersey 07632 isbn = 013666959X

References

[1] Belshaw, Christopher (2001). *Environmental Philosophy*. Chesham: Acumen. ISBN 1902683218.

[2] " International Association of Environmental Philosophy (http://www.environmentalphilosophy.org/)". . Retrieved 2008-07-30.

Environmental law

Environmental law is a complex and interlocking body of treaties, conventions, statutes, regulations, and common law that, very broadly, operate to regulate the interaction of humanity and the rest of the biophysical or natural environment, toward the purpose of reducing the impacts of human activity, both on the natural environment and on humanity itself. The topic may be divided into two major areas: (1) pollution control and remediation, and (2) resource conservation and management. Laws dealing with pollution are often media-limited - i.e., pertain only to a single environmental medium, such as air, water (whether surface water, groundwater or oceans), soil, etc. - and control both emissions of pollutants into the medium, as well as liability for exceeding permitted emissions and responsibility for cleanup. Laws regarding resource conservation and management generally focus on a single resource - e.g., natural resources such as forests,

Water, air and the byproducts of human activity do not respect political boundaries.

mineral deposits or animal species, or more intangible resources such as especially scenic areas or sites of high archeological value - and provide guidelines for and limitations on the conservation, disturbance and use of those resources. These areas are not mutually exclusive - for example, laws governing water pollution in lakes and rivers may also conserve the recreational value of such water bodies. Furthermore, many laws that are not exclusively "environmental" nonetheless include significant environmental components and integrate environmental policy decisions. Municipal, state and national laws regarding development, land use and infrastructure are examples.

Environmental law draws from and is influenced by principles of → environmentalism, including → ecology, → conservation, stewardship, responsibility and sustainability. Pollution control laws generally are intended (often with varying degrees of emphasis) to protect and preserve both the natural environment and human health. Resource conservation and management laws generally balance (again, often with varying degrees of emphasis) the benefits of preservation and economic exploitation of resources. From an economic perspective environmental laws may be understood as concerned with the prevention of present and future externalities, and preservation of common resources from individual exhaustion. The limitations and expenses that such laws may impose on commerce, and the often unquantifiable (non-monetized) benefit of environmental protection, have generated and continue to generate significant controversy.

Sources

Given the broad scope of environmental law, no fully definitive list of environmental laws is possible. The following discussion and resources give an indication of the breadth of law that falls within the "environmental" metric.

United States

Laws from every stratum of the laws of the United States pertain to environmental issues. The United States Congress has passed a number of landmark environmental regulatory regimes, but many other federal laws are equally important, if less comprehensive. Concurrently, the legislatures of the fifty states have passed innumerable comparable sets of laws.[1] These state and federal systems are foliated with layer upon layer of administrative regulation. Meanwhile, the U.S. judicial system reviews not only the legislative enactments, but also the administrative decisions of the many agencies dealing with environmental issues. Where the statutes and regulations end, the common law begins.

Federal Regulation

Consistent with the federal statutes that they administer, U.S. federal agencies promulgate regulations in the Code of Federal Regulations that fill out the broad programs enacted by Congress. Primary among these is Title 40 of the Code of Federal Regulations, containing the regulations of the Environmental Protection Agency. Other import CFR sections include Title 10 (energy), Title 18 (Conservation of Power and Water Resources), Title 21 (Food and Drugs), Title 33 (Navigable Waters), Title 36 (Parks, Forests and Public Property), Title 43 (Public Lands: Interior) and Title 50 (Wildlife and Fisheries).

Judicial Decisions

The federal and state judiciaries have played an important role in the development of environmental law in the United States, in many cases resolving significant controversy regarding the application of federal environmental laws in favor of environmental interests. The decisions of the Supreme Court in cases such as Calvert Cliffs Coordinating Committee v. U.S. Atomic Energy Commission (broadly reading the procedural requirements of the National Environmental Policy Act), Tennessee Valley Authority v. Hill (broadly reading the Endangered Species Act), and, much more recently, Massachusetts v. EPA (requiring EPA to regulate greenhouse gases under the Clean Air Act) have had policy impacts far beyond the facts of the particular case.

Common Law

The common law of tort is an important tool for the resolution of environmental disputes that fall beyond the confines of regulated activity. Prior to the modern proliferation of environmental regulation, the doctrines of nuisance, trespass, negligence, and strict liability apportioned harm and assigned liability for activities that today would be considered pollution and likely governed by regulatory regimes.[2] These doctrines remain relevant, and most recently have been used by plaintiffs seeking to impose liability for the consequences of global climate change.[3] .

The common law also continues to play a leading role in American water law, in the doctrines of riparian rights and prior appropriation.

Administration

United States

In the United States, responsibilities for the administration of environmental laws are divided between numerous federal and state agencies with varying, overlapping and sometimes conflicting missions. The Environmental Protection Agency is the most well-known federal agency, with jurisdiction over many of the country's national air, water and waste and hazardous substance programs.[4] Other federal agencies, such as the U.S. Fish and Wildlife Service and National Park Service pursue primarily conservation missions,[5] while still others, such as the United States Forest Service and the Bureau of Land Management, tend to focus more on beneficial use of natural resources.[6]

Federal agencies operate within the limits of federal jurisdiction. For example, EPA's jurisdiction under the Clean Water Act is limited to "waters of the United States." Furthermore in many cases federal laws allow for more stringent regulation by states, and of transfer of certain federally-mandated responsibilities from federal to state control. U.S. state governments, therefore, administering state law adopted under state police powers or federal law by delegation, uniformly include environmental agencies.[7] The extent to which state environmental laws are based on or depart from federal law varies from jurisdiction to jurisdiction.

Thus, while a permit to fill non-federal wetlands might require a permit from a single state agency, larger and more complex endeavors - for example, the construction of a coal-fired power plant - might require approvals from numerous federal and state agencies.

Enforcement

United States

In the United States, violations of environmental laws are generally civil offenses, resulting in monetary penalties and, perhaps, civil sanctions such as injunction. Many environmental laws do, however, provide for criminal penalties for egregious violations. Often, environmental agencies include separate enforcement offices, with duties including monitoring permitted activities, performing compliance inspections, issuing citations and prosecuting (civilly or criminally, depending on the violation) wrongdoing. EPA's Office of Enforcement and Compliance Assurance is one such agency. Others, such as the United States Park Police, carry out more traditional law enforcement activities.

Adjudicatory proceedings for environmental violations are often handled by the agencies themselves under the strictures of administrative law. In some cases, appeals are also handled internally (for example, EPA's Environmental Appeals Board). Generally, final agency determinations may subsequently be appealed to the appropriate court.

Controversy

Necessity

The necessity of directly regulating a particular activity due to the activity's environmental consequences is often a subject of debate. These debates may be scientific - for example, scientific uncertainty undergirds the ongoing debate over greenhouse gas regulation, and is a major factor in the debate over whether to ban pesticides.[8]

Cost

It is very common for regulated industry to argue against environmental regulation on the basis of cost. Indeed, in the U.S. estimates of the environmental regulation's total costs reach 2% of GDP,[9] and any new regulation will arguably contribute in some way to that burden. Difficulties arise, however, in performing cost-benefit analysis. The value of a healthy ecosystem is not easily quantified, nor the value of clean air, species diversity, etc. Furthermore environmental issues may gain an → ethical or moral dimension that would discounts cost.

Effectiveness

Environmental interests will often criticize environmental regulation as inadequately protective of the environment. Furthermore, strong environmental laws do not guarantee strong enforcement.

Education

Environmental law courses are offered as elective courses in the second and third years of JD study at many American law schools. Curricula vary: an introductory course might focus on the "big five" federal statutes - NEPA, CAA, CWA, CERCLA and RCRA (or FIFRA) - and may be offered in conjunction with a natural resources law course. Smaller seminars mights be offered on more focused topics. Some U.S. law schools also offer an LLM or JSD specialization in environmental law. Additionally, several law schools host legal clinics that focus on environmental law, providing students with an opportunity to learn about environmental law in the context of real world disputes involving actual clients.[10] U.S. News & World Report has consistently ranked University of Oregon School of Law, Vermont Law School, Lewis & Clark Law School, Pace University School of Law, Tulane University School of Law, and Georgetown University Law Center as among the best Environmental Law programs in the United States.[11]

Many law schools host student-published law journals. The environmental law reviews at Yale, Harvard, Stanford, Columbia and NYU law schools are regularly the most-cited such publications. [12]

The IUCN Academy of Environmental Law[13] is a network of some 60 law schools worldwide that specialise in the research and teaching of environmental law.

See also

- Citizen suit
- Clean tech law
- Energy law
- Earth jurisprudence
- Environmental Bill of Rights (EBR)
- Environmental compensation
- Environmental contract
- Environmental crime
- Environmental criminology
- Environmental economics
- environmental engineering law
- Environmental justice
- Environmental Law (Law Review)
- Environmental tariff
- Environmental vandalism

- International environmental law
- *Journal of Land Use and Environmental Law*
- List of environmental agreements
- List of environmental lawsuits
- List of ministers of the environment

- Right to environmental protection
- Timeline of major U.S. environmental and occupational health regulation
- Toxic tort
- United States environmental law
- War and environmental law

References and bibliography

[1] See, e.g., Pennsylvania (http://www.dep.state.pa.us/dep/SUBJECT/LEGSREGS/laws.htm)

[2] See West's Encyclopedia of American Law, Environmental Law (http://www.answers.com/topic/environmental-law).

[3] Cases collected at Climatecasechart.com (Common Law Claims) (http://www.climatecasechart.com/)

[4] See EPA, Laws That We Administer (http://www.epa.gov/lawsregs/laws/index.html); EPA, Alphabetical Listing of EPA programs (http://www.epa.gov/epahome/abcpgram.htm).

[5] See National Park Service, Our Mission (http://www.nps.gov/legacy/mission.html); USFWS, National Policy Issuance #99-01 (http://www.fws.gov/policy/npi99_01.html)

[6] See USFS Mission Statement (http://www.fs.fed.us/aboutus/mission.shtml).

[7] See EPA, State Environmental Agencies (http://www.epa.gov/epahome/state.htm)

[8] See, e.g., DDT.

[9] Pizer & Kopp, Calculating the Costs of Environmental Regulation, 1 (2003 Resources for the Future) (http://www.rff.org/Documents/RFF-DP-03-06.pdf).

[10] See, e.g., Adam Babich, The Apolitical Law School Clinic, 11 Clinical L. Rev. 447 (2005).

[11] US News & World Report (http://grad-schools.usnews.rankingsandreviews.com/grad/law/environ)

[12] See Washington and Lee University School of Law, Law Journal Rankings (http://lawlib.wlu.edu/lj/index.aspx)

[13] IUCN Academy of Environmental Law (http://www.iucnael.org)

Bibliography

- Menell, P.S. (ed), *Environmental Law* (Ashgate Publishing, Burlington, 2003).
- Clayton, Susan. 2000. Models of justice in the environmental debate. Journal of Social Issues 56 (3): 459 – 474.
- Croci, E. (ed), *The Handbook of Environmental Voluntary Agreements* (Springer, New York, 2005).
- Freeman, J. and Kolstad, C.D. (eds), *Moving to Markets in Environmental Regulation* (Oxford University Press, New York, 2006).
- Lans, C. 2007. Politically incorrect and bourgeois: Nariva Swamp is sufficient onto itself. Lulu.com.
- Richardson, B.J. and S. Wood (eds), *Environmental Law for Sustainability* (Hart Publishing, Oxford, 2006) (http://www.hartpub.co.uk/books/details.asp?sc=1-84113-544-5)
- Saxe, D., "Environmental Offences: Corporate Responsibility and Executive Liability" (Canada Law Book, Aurora, 1990).
- Stone, Christopher, D. 1974. Should trees have legal standing? Toward legal rights for natural objects. California: William Kaufman, Inc.

External links

- American Bar Association Section of Environment, Energy and Resources (http://www.abanet.org/environ/)
- U.S. Environmental Protection Agency (http://www.epa.gov/)
- Environmental Law Institute (ELI) (http://www.eli.org/)
- Centre for International Environmental Law (http://www.ciel.org/)
- International Environmental Law Research Centre (IELRC) (http://www.ielrc.org/)

Environmental sociology

Environmental sociology is typically defined as the sociological study of societal-environmental interactions, although this definition immediately presents the perhaps insolvable problem of separating human cultures from the rest of the environment. Although the focus of the field is the relationship between society and environment in general, environmental sociologists typically place special emphasis on studying the social factors that cause environmental problems, the societal impacts of those problems, and efforts to solve the problems. In addition, considerable attention is paid to the social processes by which certain environmental conditions become socially defined as problems.

Although there was sometimes acrimonious debate between the constructivist and realist "camps" within environmental sociology in the 1990s, the two sides have found considerable common ground as both increasingly accept that while most environmental problems have a material reality they nonetheless become known only via human processes such as scientific knowledge, activists' efforts, and media attention. In other words, most environmental problems have a real ontological status despite our knowledge/awareness of them stemming from social processes, processes by which various conditions are constructed as problems by scientists, activists, media and other social actors. Correspondingly, environmental problems must all be understood via social processes, despite any material basis they may have external to humans. This interactiveness is now broadly accepted, but many aspects of the debate continue in contemporary research in the field.

History

Modern thought surrounding human-environment relations is traced back to Charles Darwin. Darwin's concept of natural selection suggested that certain social characteristics played a key role in the survivability of groups in the natural environment. Although typically taken at the micro level, evolutionary principles, particularly adaptability, serve as a microcosm of human ecology. Work by Humphrey and Buttel (2002) traces the linkages between Darwin's work on natural selection, human ecological sociology, and environmental sociology.

Academic

It became recognized in the latter half of the 20th century that biological determinism failed to fully explain the relationship between humans and the environment. As the application of social determinism became more useful, the role of sociology became more pervasive in analyzing environmental conditions. At first, classical sociology saw social and cultural factors as the only cause of other social and cultural conditions. This lens ignored the concept of environmental determinism or the environmental factors that cause social phenomena.

The works of William R. Catton, Jr. and Riley Dunlap challenged the constricted → anthropocentrism of classical sociology. In the late 1970s, they called for a new holistic, or systems perspective. Since the 1970s, sociology has noticeably transformed to include environmental forces in social explanations. Environmental sociology emerged as a coherent subfield of inquiry after the → environmental movement of the 1960s and early 1970s. It has now solidified as a respected, interdisciplinary subject in academia.

Concepts

Existential dualism

The duality of the human condition rests with cultural uniqueness and evolutionary traits. From one perspective, humans are embedded in the ecosphere and coevolved alongside other species. Humans share the same basic ecological dependencies as other inhabitants of nature. From the other perspective, humans are distinguished from other species because of their innovative capacities, distinct cultures and varied institutions. Human creations have the power to independently manipulate, destroy, and transcend the limits of the natural environment (Buttel and Humphrey, 2002: p.47).

Support for each perspective varies among different communities. Biologists and ecologists typically put more weight on the first perspective. Social scientists, on the other hand, emphasize the second perspective. This division has shaped the foundation for the primary paradigms of environmental sociology.

Societal-environmental dialectic

In 1975, the highly influential work of Allan Schnaiberg transfigured environmental sociology, proposing a societal-environmental dialectic. This conflictual concept has overwhelming political salience. First, the economic synthesis states that the desire for economic expansion will prevail over ecological concerns. Policy will decide to maximize immediate economic growth at the expense of environmental disruption. Secondly, the managed scarcity synthesis concludes that governments will attempt to control only the most dire of environmental problems to prevent health and economic disasters. This will give the appearance that governments act more environmentally conscious than they really do. Tertiary, the ecological synthesis generates a hypothetical case where environmental degradation is so severe that political forces would respond with sustainable policies. The driving factor would be economic damage caused by environmental degradation. The economic engine would be based on renewable resources at this point. Production and consumption methods would adhere to sustainability regulations.

These conflict-based syntheses have several potential outcomes. One is that the most powerful economic and political forces will preserve the status quo and bolster their dominance. Historically, this is the most common occurrence. Another potential outcome is for contending powerful parties to fall into a stalemate. Lastly, tumultuous social events may result that redistribute economic and political resources.

Treadmill of production

In 1980, Schnaiberg developed a conflict theory on human-environment interaction. The theory is that capitalism is driven by higher profitability and thereby must continue to grow and attract investments to survive in a competitive market. This identifies the imperative for continued economic growth levels that, once achieved, accelerate the need for future growth. This growth in production requires a corresponding growth in consumption. The process contains a chief paradox; economic growth is socially desired but environmental degradation is a common consequence that in turn disrupts long-run economic expansion (Schnaiberg 1980).

Paradigms

Human Exemptionalism Paradigm (HEP)

The HEP theory claims that humans are such a uniquely superior species that they are exempt from environmental forces. Shaped by the leading Western worldview of the time, this was the popular societal paradigm from the industrial revolution until the second half of the 20th century. Human dominance was justified by the uniqueness of culture, which is far more adaptable than biological traits. Culture also has the capacity to accumulate and innovate, making it an unbounded resource capable of solving all natural problems. As humans are not governed by natural conditions, they have complete control of their own destiny. Any potential limitation posed by the natural world is surpassable using human ingenuity.

New Ecological Paradigm (NEP)

In the 1970s, scholars began recognizing the limits of what would be termed the Human Exemptionalism Paradigm. Catton and Dunlap suggested a new perspective that took environmental variables into full account. They coined a new theory, the New Ecological Paradigm, with assumptions contrary to the HEP. The NEP recognizes the innovative capacity of humans, but that says humans are still ecologically interdependent as with other species. The NEP notes the power of social and cultural forces but does not profess social determinism. Instead, humans are impacted by the cause, effect, and feedback loops of ecosystems. The earth has a finite level of natural resources and waste repositories. Thus, the biophysical environment can impose restraints on human activity.

Events

Modern environmentalism

The 1960s built strong cultural momentum for environmental causes, giving birth to the modern → environmental movement. Widespread green consciousness moved vertically within society, resulting in a series of federal policy changes in the 1970s. This period was known as the "Environmental Decade" with the creation of the United States Environmental Protection Agency and passing of the Endangered Species Act, Clean Water Act, and amendments to the Clean Air Act. → Earth Day of 1970, celebrated by millions of participants, represented the modern age of environmental thought. The → environmental movement continued with incidences such as Love Canal.

Historical studies

While the current mode of thought expressed in environmental sociology was not prevalent until the age of modernity, its application is now used in analysis of ancient peoples. Societies including Easter Island, the Anaszi, and the Mayans ended abruptly, largely due to poor environmental management. The collapse of the Mayans sent a historic message that even advanced cultures are vulnerable to ecological suicide. At the same time, societal successes include New Guinea, Tikopia island, and Japan, whose inhabitants have lived sustainably for 46,000 years.

See also

- Ecological anthropology
- Agroecology
- Ecological modernization theory
- Environmental design
- Environmental design and planning

- Environmental Economics
- Environmental Policy
- → Environmentalism
- Human ecology

- Important publications in environmental sociology
- Sociology of architecture
- United States Environmental Protection Agency

References

- Buttel, Frederick H. and Craig R. Humphrey. 2002. "Sociological Theory and the Natural Environment." Pp. 33-69 in *Handbook of Environmental Sociology* edited by Riley E. Dunlap and William Michelson, Westport, CT: Greenwood Press.
- Diamond, Jared. (2005) *Collapse: How Societies Choose to Fail or Succeed*. New York: Viking. ISBN 0-670-03337-5.
- Dunlap, Riley E., Frederick H. Buttel, Peter Dickens, and August Gijswijt (eds.) 2002. *Sociological Theory and the Environment: Classical Foundations, Contemporary Insights* (Rowman & Littlefield, ISBN 0-7425-0186-8).
- Dunlap, Riley E., and William Michelson (eds.) 2002. *Handbook of Environmental Sociology* (Greenwood Press, ISBN 0-313-26808-8)

- Freudenburg, William R., and Robert Gramling. "The Emergence of Environmental Sociology: Contributions of Riley E. Dunlap and William R. Catton, Jr.", *Sociological Inquiry* 59(4): 439-452
- Harper, Charles. 2004. *Environment and Society: Human Perspectives on Environmental Issues*. Upper Saddle River, New Jersey: Pearson Education, Inc. ISBN 0131113410
- Humphrey, Craig R., and Frederick H. Buttel. 1982. *Environment, Energy, and Society*. Belmont, California: Wadsworth Publishing Company. ISBN 0-534-00964-6
- Humphrey, Craig R., Tammy L. Lewis and Frederick H. Buttel. 2002. *Environment, Energy and Society: A New Synthesis*. Belmont, California: Wadsworth/Thompson Learning. ISBN 0-534-57955-8
- Mehta, Michael, and Eric Ouellet. 1995. *Environmental Sociology: Theory and Practice*, Toronto: Captus Press.
- Redclift, Michael, and Graham Woodgate, eds. 1997. *International Handbook of Environmental Sociology* (Edgar Elgar, 1997; ISBN 1-84064-243-2)
- Schnaiberg, Allan. 1980. *The Environment: From Surplus to Scarcity*. New York: Oxford University Press. Available: http://media.northwestern.edu/sociology/schnaiberg/1543029_environmentsociety/index.html.
- http://www.socialresearchmethods.net/Gallery/Neto/Envsoc1.html

External links

- ASA Section on Environment and Technology [1]
- Environmental sociology [2] - a resource page.
- ISA Research Committee on Environment and Society (RC24) [3]
- Ecology and Society book [4]

References

[1] http://envirosoc.org/
[2] http://www.socialresearchmethods.net/Gallery/Neto/Envsoc1.html
[3] http://www.isa-sociology.org/rc24.htm
[4] http://www.sussex.ac.uk/Users/ssfa2/ecology.html

Ecotheology

Ecotheology is a form of constructive theology that focuses on the interrelationships of religion and nature, particularly in the light of environmental concerns. Ecotheology generally starts from the premise that a relationship exists between human religious/spiritual worldviews and the degradation of nature. It explores the interaction between ecological values, such as sustainability, and the human domination of nature. The movement has produced numerous religious-environmental projects around the world.

The burgeoning awareness of environmental crisis has led to widespread religious reflection on the human relationship with the earth. Such reflection has strong precedents in most religious traditions in the realms of ethics and cosmology, and can be seen as a subset or corollary to the theology of nature. Christian ecotheology draws on the writings of such authors as Jesuit priest and paleontologist Pierre Teilhard de Chardin, process theologian Alfred North Whitehead, and is well-represented in Protestantism by John Cobb, Jr. and Jurgen Moltmann and ecofeminist theologians Rosemary Radford Ruether, Catherine Keller and Sallie McFague. Creation theology is another important expression of ecotheology that has been developed and popularized by Matthew Fox, the former Catholic priest. Abraham Joshua Heschel and Martin Buber, both Jewish theologians, have also left their mark on Christian ecotheology, and provide significant inspiration for Jewish ecotheology.

Hindu ecotheology includes writers such as Vandana Shiva. Seyyid Hossein Nasr, a liberal Muslim theologian, was one of the earlier voices calling for a re-evaluation of the Western relationship to nature.

Precedents in religious thought

Christianity has often been viewed as the source of negative values towards the environment (see below), but there are many voices within the Christian tradition whose vision embraces the well-being of the earth and all creatures. While St. Francis of Assisi is one of the more obvious influences on Christian ecotheology, there are many theologians and teachers whose work has profound implications for Christian thinkers. Many of these are less well-known in the West because their primary influence has been on the Orthodox Church rather than the Roman Catholic Church.

The significance of indigenous traditions for the development of ecotheology can also not be understated.

Background

The relationship of theology to the modern ecological crisis became an intense issue of debate in Western academia in 1967, following the publication of the article, "The Historical Roots of Our Ecological Crisis, " by Lynn White, Jr., Professor of History at the University of California at Los Angeles. In this work, White puts forward a theory that the Christian model of human dominion over nature has led to environmental devastation.

In 1973, theologian Jack Rogers published an article in which he surveyed the published studies of approximately twelve theologians which had appeared since White's article. They reflect the search for "an appropriate theological model" which adequately assesses the biblical data regarding any relationship of God, humans, and nature.

Further exploration

Elisabet Sahtouris is a biologist who promotes a vision she believes will result in the sustainable health and well-being of humanity within the larger living systems of Earth and the cosmos. She is a lecturer in Gaia Theory and a co-worker with James Lovelock and Lynn Margulis.

Annie Dillard, Pulitzer Prize-winning American author, also combined observations on nature and philosophical explorations in several ecotheological writings, including *Pilgrim at Tinker Creek*.

Valerie Brown is a science and environmental journalist based in Portland, Oregon, whose work has appeared in *Environmental Health Perspectives, 21stC*, and other publications. She writes regularly about ecotheology.

Terry Tempest Williams is a Mormon writer who sensitively and imaginatively explores ecotheology in her very personal writing.

The majority of the content of *Indians of the Americas*, by former Bureau of Indian Affairs head John Collier, concerns the link between ecological sustainability and religion among Native North and South Americans.

See also

- Stewardship (theology)
- Faith in Place
- Human ecology
- Spiritual ecology
- Religion and ecology

References

- Rogers, J. (1973). "Ecological Theology: The Search for an Appropriate Theological Model." Reprinted from *Septuagesino Anno: Theologiche Opstellen Aangebsden Aan Prof. Dr. G. C. Berkower*. The Netherlands: J.H. Kok.
- White, L. Jr. (1971). "The Historical Roots of our Ecologic Crisis." Reprinted in A.E. Lugo & S.C. Snedaker (Eds.) *Readings on Ecological Systems: Their Function and Relation to Man*. New York: MSS Educational Publishing.
- *"Why Care for Earth's Environment?"* (in the series *"The Bible's Viewpoint"*) is a two-page article in the December 2007 issue of Awake! magazine. This represents the Bible's viewpoint according to the viewpoint of Jehovah's Witnesses.

External links

- MarvelBelieveCare.org [1] provides free online educational materials about the Bible and caring for God's creation
- ARC - Alliance of Religions and Conservation (Bath UK) [2]
- CCC - Catholic Conservation Center (Wading River NY US) [3]
- CofDE - Church of Deep Ecology (Minneapolis MN USA) [4]
- COEJL - Coalition on the Environment and Jewish Life (NYC US) [5]
- CRLE - Center for Respect of Life and Environment (Washington DC US) [6]
- EEN - Evangelical Environmental Network (Suwanee Ga US) [7]
- EJP - Environmental Justice Program (USCCB (U.S. Conference of Catholic Bishops) SDWP, Washington DC US) [8]
- The Forum on Religion and Ecology (Harvard University, Cambridge Ma US) [9]

- ISSRNC - International Society for the Study of Religion, Nature, and Culture (Dept. of Religion, Univ. of Florida, Gainesville FL US) [10]
- NCC - Eco-Justice Program (Natl Council of Churches of Christ, Washington DC US) [11]
- NRPE - National Religious Partnership for the Environment (Amherst Ma US) [12]
- Web of Creation (Lutheran School of Theology, Chicago IL US) [13]
- Christians' Ecological Responsibility [14]
- The Ecotheology of Annie Dillard [15] Annie Dillard -
- The rise of ecotheology [16]
- Ecotheology: The Journal of Religion, Nature and the Environment [17]

References

[1] http://www.MarvelBelieveCare.org

[2] http://www.arcworld.org/

[3] http://conservation.catholic.org/

[4] http://www.churchofdeepecology.org/

[5] http://www.coejl.org/index.php

[6] http://www.crle.org/

[7] http://www.creationcare.org/

[8] http://www.usccb.org/sdwp/ejp/

[9] http://environment.harvard.edu/religion/main.html

[10] http://www.religionandnature.com/society/

[11] http://www.nccecojustice.org/

[12] http://www.nrpe.org/

[13] http://www.webofcreation.org/

[14] http://www.asa3.org/ASA/PSCF/1993/PSCF3-93Stanton.html

[15] http://www.crosscurrents.org/dillard.htm

[16] http://www.columbia.edu/cu/21stC/issue-3.4/brown.html

[17] http://www.equinoxpub.com/journals/main.asp?jref=6

Ecological economics

Ecological economics is a transdisciplinary field of academic research that aims to address the interdependence and coevolution of human economies and natural ecosystems over time and space.[2] It is distinguished from environmental economics, which is the mainstream economic analysis of the environment, by its treatment of the economy as a subsystem of the ecosystem and its emphasis upon preserving natural capital.[3] One survey of German economists found that ecological and environmental economics are different schools of economic thought, with ecological economists emphasizing "strong" sustainability and rejecting the proposition that natural capital can be substituted for human-made capital.[4]

Ecological economics was founded in the works of Kenneth E. Boulding, Nicholas Georgescu-Roegen, Herman Daly, Robert Costanza, and others. The related field of green economics is, in general, a more politically applied form of the subject.[5] [6]

The identity of ecological economics as a field has been described as fragile, with no generally accepted theoretical framework and a knowledge structure which is not clearly defined.[7] According to ecological economist Malte Faber, ecological economics is defined by its focus on nature, justice, and time. Issues of intergenerational equity, irreversibility of environmental change, uncertainty of long-term

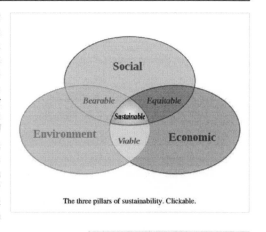

The three pillars of sustainability. Clickable.

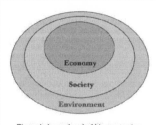

Three circles enclosed within one another showing how both economy and society are subsets of our planetary ecological system. This view is useful for correcting the misconception, sometimes drawn from the previous "three pillars" diagram, that portions of social and economic systems can exist independently from the environment.[1]

outcomes, and sustainable development guide ecological economic analysis and valuation.[7] Ecological economists have questioned fundamental mainstream economic approaches such as cost-benefit analysis, and the separability of economic values from scientific research, contending that economics is unavoidably normative rather than positive (empirical).[8] Positional analysis, which attempts to incorporate time and justice issues, is proposed as an alternative.[9] [10]

Ecological economics includes the study of the metabolism of society, that is, the study of the flows of energy and materials that enter and exit the economic system. This subfield is also called biophysical economics, sometimes referred to also as bioeconomics. It is based on a conceptual model of the economy connected to, and sustained by, a flow of energy, materials, and ecosystem services. Analysts from a variety of disciplines have conducted research on the economy-environment relationship, with concern for energy and material flows and sustainability, environmental quality, and economic development.

Nature and ecology

A simple circular flow of income diagram is replaced in ecological economics by a more complex flow diagram reflecting the input of solar energy, which sustains natural inputs and environmental services which are then used as units of production. Once consumed, natural inputs pass out of the economy as pollution and waste. The potential of an environment to provide services and materials is referred to as an "environment's source function", and this function is depleted as resources are consumed or pollution contaminates the resources. The "sink function" describes an environment's ability to absorb and render harmless waste and pollution: when waste output exceeds the limit of the sink function, long-term damage

Environmental Scientist sampling water.

occurs.[11] :8 Some persistent pollutants, such as some organic pollutants and nuclear waste are absorbed very slowly or not at all; ecological economists emphasize minimizing "cumulative pollutants".[11] :28 Pollutants affect human health and the health of the climate.

The economic value of natural capital and ecosystem services is accepted by mainstream environmental economics, but is emphasized as especially important in ecological economics. Ecological economists may begin by estimating how to maintain a stable environment before assessing the cost in dollar terms.[11] :9 Ecological economist Robert Costanza led an attempted valuation of the global ecosystem in 1997. Initially published in *Nature*, the article concluded on $33 trillion with a range from $16 trillion to $54 trillion (in 1997, total global GDP was $27 trillion).[12] Half of the value went to nutrient cycling. The open oceans, continental shelves, and estuaries had the highest total value, and the highest per-hectare values went to estuaries, swamps/floodplains, and seagrass/algae beds. The work was criticized by articles in *Ecological Economics* Volume 25, Issue 1, but the critics acknowledged the positive potential for economic valuation of the global ecosystem.[11] :129

The Earth's carrying capacity is another central question. This was first examined by Thomas Malthus, and more recently in an MIT study entitled *Limits to Growth*. Although the predictions of Malthus have not come to pass, some limit to the Earth's ability to support life are acknowledged. In addition, for real GDP per capita to increase real GDP must increase faster than population growth. Diminishing returns suggest that productivity increases will slow if major technological progress is not made. Food production may become a problem, as erosion, an impending water crisis, and soil salinity (from irrigation) reduce the productivity of agriculture. Ecological economists argue that industrial agriculture, which exacerbates these problems, is not sustainable agriculture, and are generally inclined favorably to organic farming, which also reduces the output of carbon.[11] :26 Global wild fisheries are believed to have peaked and begun a decline, with valuable habitat such as estuaries in critical condition.[11] :28 The aquaculture or farming of piscivorous fish, like salmon, does not help solve the problem because they need to be feed products from other fish. Studies have shown that salmon farming has major negative impacts on wild salmon, as well as the forage fish that need to be caught to feed them.[13] [14] Since animals are higher on the trophic level, they are less efficient sources of food energy. Reduced consumption of meat would reduce the demand for food, but as nations develop, they adopt high-meat diets similar to the United States. Genetically modified food (GMF) a conventional solution to the problem, have problems – Bt corn produces its own Bacillus thuringiensis, but the pest resistance is believed to be only a matter of time.[11] :31 The overall effect of GMF on yields is contentious, with the USDA and FAO acknowledging that GMFs do not necessarily have higher yields and may even have reduced yields.[15]

Global warming is now widely acknowledged as a major issue, with all national scientific academies expressing agreement on the importance of the issue. As the population growth intensifies and energy demand increases, the world faces an energy crisis. Some economists and scientists forecast a global ecological crisis if energy use is not contained – the Stern report is an example. The disagreement has sparked a vigorous debate on issue of discounting and intergenerational equity.

GLOBAL GEOCHEMICAL CYCLES CRITICAL FOR LIFE

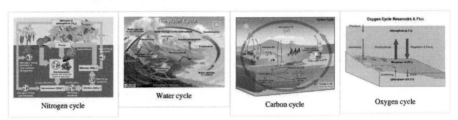

Nitrogen cycle   Water cycle   Carbon cycle   Oxygen cycle

Ethics

Mainstream economics has attempted to become a value-free 'hard science', but ecological economists argue that value-free economics is generally not realistic. Ecological economics is more willing to entertain alternative conceptions of utility, efficiency, and cost-benefits such as positional analysis or multi-criteria analysis. Ecological economics is typically viewed as economics for sustainable development,[16] and may have goals similar to green politics.

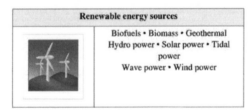

Renewable energy sources
Biofuels • Biomass • Geothermal Hydro power • Solar power • Tidal power Wave power • Wind power

Schools of thought

Various competing schools of thought exist in the field. Some are close to resource and environmental economics while others are far more heterodox in outlook. An example of the latter is the *European Society for Ecological Economics*. An example of the former is the Swedish *Beijer International Institute of Ecological Economics*.

Differentiation from mainstream schools

In ecological economics, natural capital is added to the typical capital asset analysis of land, labor, and financial capital. Ecological economics uses tools from mathematical economics, but may apply them more closely to the natural world. Whereas mainstream economists tend to be technological optimists, ecological economists are inclined to be technological pessimists. They reason that the natural world has a limited carrying capacity and that its resources may run out. Since destruction of important environmental resources could be practically irreversible and catastrophic, ecological economists are inclined to justify cautionary measures based on the precautionary principle.[17]

The most cogent example of how the different theories treat similar assets is tropical rainforest ecosystems, most obviously the Yasuni region of Ecuador. While this area has substantial deposits of bitumen it is also one of the most diverse ecosystems on Earth and some estimates establish it has over 200 undiscovered medical substances in its genomes - most of which would be destroyed by logging the forest or mining the bitumen. Effectively, the instructional capital of the genomes is undervalued by both classical and neoclassical means which would view the rainforest primarily as a source of wood, oil/tar and perhaps food. Increasingly the carbon credit for leaving the extremely carbon-intensive ("dirty") bitumen in the ground is also valued - the government of Ecuador set a price of US$350M for an oil lease with the intent of selling it to someone committed to never exercising it at all and instead preserving the rainforest. Bill Clinton, Paul Martin and other former world leaders have become closely involved in this project which includes lobbying for the issue of International Monetary Fund Special Drawing Rights to recognize the rainforest's value directly within the framework of the Bretton Woods institutions. If successful this would be a major victory for advocates of ecological economics as the new mainstream form of economics.

History and Development

Early interest in ecology and economics dates back to the 1960s and the work by Kenneth Boulding and Herman Daly, but the first meetings occurred in the 1980s. It began with a 1982 symposium in Sweden [18] which was attended by people who would later be instrumental in the field, including Robert Costanza, Herman Daly, Charles Hall, Ann-Mari Jansson, Bruce Hannon, H.T. Odum, and David Pimentel. Most were ecosystem ecologists or mainstream environmental economists, with the exception of Daly. In 1987, Daly and Costanza edited an issue of *Ecological Modeling* to test the waters. *Ecological Economics* by Juan Martinez-Alier [19] was published later that year.[20]

European conceptual founders include Nicholas Georgescu-Roegen (1971), William Kapp (1944)[21] and Karl Polanyi (1950).[22] Some key concepts of what is now ecological economics are evident in the writings of E.F. Schumacher, whose book *Small Is Beautiful – A Study of Economics as if People Mattered* (1973) was published just a few years before the first edition of Herman Daly's comprehensive and persuasive *Steady-State Economics* (1977).[23] [24] Other figures include ecologists C.S. Holling, H.T. Odum and Robert Costanza, biologist Gretchen Daily and physicist Robert Ayres. CUNY geography professor David Harvey explicitly added ecological concerns to political economic literature. This parallel development in political economy has been continued by analysts such as sociologist John Bellamy Foster.

The antecedents can be traced back to the Romantics of the 1800s as well as some Enlightenment political economists of that era. Concerns over population were expressed by Thomas Malthus, while John Stuart Mill hypothesized that the "stationary state" of an economy might be something that could be considered desirable, anticipating later insights of modern ecological economists, without having had their experience of the social and ecological costs of the dramatic post-World War II industrial expansion. As Martinez-Alier explores in his book the debate on energy in economic systems can also be traced into the 1800s e.g. Nobel prize-winning chemist, Frederick Soddy (1877-1956). Soddy criticized the prevailing belief of the economy as a perpetual motion machine, capable of generating infinite wealth — a criticism echoed by his intellectual heirs in the now emergent field of ecological economics.[25]

The Romanian economist Nicholas Georgescu-Roegen (1906-1994), who was among Daly's teachers at Vanderbilt University, provided ecological economics with a modern conceptual framework based on the material and energy flows of economic production and consumption. His *magnum opus*, *The Entropy Law and the Economic Process* (1971), has been highly influential.[26]

Articles by Inge Ropke (2004, 2005)[27] and Clive Spash (1999)[28] cover the development and modern history of ecological economics and explain its differentiation from resource and environmental economics, as well as some of the controversy between American and European schools of thought. An article by Robert Costanza, David Stern, Lining He, and Chunbo Ma [29] responded to a call by Mick Common to determine the foundational literature of ecological economics by using citation analysis to examine which books and articles have had the most influence on the development of the field.

Topics

Methodology

Thermodynamic equations
Laws of thermodynamics
Zeroth law
First law
Second law
Third law
Conjugate variables
Thermodynamic potential
Material properties
Maxwell relations
Bridgman's equations
Exact differential
Table of thermodynamic equations
[30]

The primary objective of ecological economics (EE) is to ground economic thinking and practice in physical reality, especially in the laws of physics (particularly the laws of thermodynamics) and in knowledge of biological systems. It accepts as a goal the improvement of human well-being through development, and seeks to ensure achievement of this through planning for the sustainable development of ecosystems and societies. Of course the terms development and sustainable development are far from lacking controversy. Richard Norgaard argues traditional economics has hi-jacked the development terminology in his book *Development Betrayed*.[31] Well-being in ecological economics is also differentiated from welfare as found in mainstream economics and the 'new welfare economics' from the 1930s which informs resource and environmental economics. This entails a limited preference utilitarian conception of value i.e., Nature is valuable to our economies, that is because people will pay for its services such as clean air, clean water, encounters with wilderness, etc.

Ecological economics distinguishes itself from neoclassical economics primarily by its assertion that the economy is an embedded within an environmental system. Ecology deals with the energy and matter transactions of life and the Earth, and the human economy is by definition contained within this system. Ecological economists feel neoclassical economics has ignored the environment, at best relegating it to be a subset of the human economy.

However, this belief disagrees with much of what the natural sciences have learned about the world, and, according to Ecological Economics, completely ignores the contributions of Nature to the creation of wealth e.g., the planetary endowment of scarce matter and energy, along with the complex and biologically diverse ecosystems that provide goods and ecosystem services directly to human communities: micro- and macro-climate regulation, water recycling, water purification, storm water regulation, waste absorption, food and medicine production, pollination, protection from solar and cosmic radiation, the view of a starry night sky, etc.

There has then been a move to regard such things as natural capital and ecosystems functions as goods and services. [32] [33] However, this is far from uncontroversial within ecology or ecological economics due to the potential for narrowing down values to those found in mainstream economics and the danger of merely regarding Nature as a commodity. This has been referred to as ecologists 'selling out on Nature'.[34] There is then a concern that ecological economics has failed to learn from the extensive literature in → environmental ethics about how to structure a plural value system.

Allocation of resources

Resource and neoclassical economics focus primarily on the efficient allocation of resources, and less on two other fundamental economic problems which are central to ecological economics: distribution (equity) and the scale of the economy relative to the ecosystems upon which it is reliant.[35] Ecological Economics also makes a clear distinction between growth (quantitative) and development (qualitative improvement of the quality of life) while arguing that neoclassical economics confuses the two. Ecological economics challenges the common normative approach taken towards natural resources, claiming that it misvalues nature by displaying it as interchangeable with human capital--labor and technology. EE counters this convention by asserting that human capital is instead complementary to and dependent upon natural systems, as human capital inevitably derives from natural systems. From these premises, it follows that economic policy has a fiduciary responsibility to the greater ecological world, and that, by misvaluing the importance of nature, sustainable progress (as opposed to economic growth) --which is the only solution to elevating the standard of living for citizens worldwide—will not result. Furthermore, ecological economists point out that, beyond modest levels, increased per-capita consumption (the typical economic measure of "standard of living") does not necessarily lead to improvements in human well-being, while this same consumption can have harmful effects on the environment and broader societal well-being.

Energy economics

It rejects the view of energy economics that growth in the energy supply is related directly to well being, focusing instead on biodiversity and creativity - or natural capital and individual capital, in the terminology sometimes adopted to describe these economically. In practice, ecological economics focuses primarily on the key issues of uneconomic growth and quality of life. Ecological economists are inclined to acknowledge that much of what is important in human well-being is not analyzable from a strictly economic standpoint and suggests an interdisciplinary approach combining social and natural sciences as a means to address this.

Thermoeconomics is based on the proposition that the role of energy in biological evolution should be defined and understood through the second law of thermodynamics, but also in terms of such economic criteria as productivity, efficiency, and especially the costs and benefits (or profitability) of the various mechanisms for capturing and utilizing available energy to build biomass and do work.[36] [37] As a result, thermoeconomics are often discussed in the field of ecological economics, which itself is related to the fields of sustainability and sustainable development.

Exergy analysis is performed in the field of industrial ecology to use energy more efficiently.[38] The term *exergy*, was coined by Zoran Rant in 1956, but the concept was developed by J. Willard Gibbs. In recent decades, utilization of exergy has spread outside of physics and engineering to the fields of industrial ecology, ecological economics, systems ecology, and energetics.

Energy accounting and balance

An energy balance can be used to track energy through a system, and is a very useful tool for determining resource use and environmental impacts, using the First and Second laws of thermodynamics, to determine how much energy is needed at each point in a system, and in what form that energy is a cost in various environmental issues. The energy accounting system keeps track of energy in, energy out, and non-useful energy versus work done, and transformations within the system. [39]

Scientists have written and speculated on different aspects of energy accounting.[40]

Environmental services

A study was carried out by Costanza and colleagues[41] to determine the 'price' of the services provided by the environment. This was determined by averaging values obtained from a range of studies conducted in very specific context and then transferring these without regard to that context. Dollar figures were averaged to a per hectare number for different types of ecosystem e.g. wetlands, oceans. A total was then produced which came out at 33 trillion US dollars (1997 values), more than twice the total GDP of the world at the time of the study. This study was criticized by pre-ecological and even some environmental economists - for being inconsistent with assumptions of financial capital valuation - and ecological economists - for being inconsistent with an ecological economics focus on biological and physical indicators.[42] . *See also ecosystem valuation and price of life.*

The whole idea of treating ecosystems as goods and services to be valued in monetary terms remains controversial to some. A common objection is that life is precious or priceless, but this demonstrably degrades to it being worthless under the assumptions of any branch of economics. Reducing human bodies to financial values is a necessary part of every branch of economics and not always in the direct terms of insurance or wages. Economics, in principle, assumes that conflict is reduced by agreeing on voluntary contractual relations and prices instead of simply fighting or coercing or tricking others into providing goods or services. In doing so, a provider agrees to surrender time and take bodily risks and other (reputation, financial) risks. Ecosystems are no different than other bodies economically except insofar as they are far less replaceable than typical labour or commodities.

Despite these issues, many ecologists and conservation biologists are pursuing ecosystem valuation. Biodiversity measures in particular appear to be the most promising way to reconcile financial and ecological values, and there are many active efforts in this regard. The growing field of biodiversity finance [43] began to emerge in 2008 in response to many specific proposals such as the Ecuadoran Yasuni proposal [44] [45] or similar ones in the Congo. US news outlets treated the stories as a "threat" [46] to "drill a park" [47] reflecting a previously dominant view that NGOs and governments had the primary responsibility to protect ecosystems. However Peter Barnes and other commentators have recently argued that a guardianship/trustee/commons model is far more effective and takes the decisions out of the political realm.

Commodification of other ecological relations as in carbon credit and direct payments to farmers to preserve ecosystem services are likewise examples that permit private parties to play more direct roles protecting biodiversity. The United Nations Food and Agriculture Organization achieved near-universal agreement in 2008 [48] that such payments directly valuing ecosystem preservation and encouraging permaculture were the only practical way out of a food crisis. The holdouts were all English-speaking countries that export GMOs and promote "free trade" agreements that facilitate their own control of the world transport network: The US, UK, Canada and Australia [49]. Increasingly the pro-GMO pro-trade view is in the extreme minority worldwide though it is disproportionately represented at the IMF, World Bank and so on.

Externalities

Ecological economics is founded upon the view that the neoclassical economics (NCE) assumption that environmental and community costs and benefits are mutually canceling *"externalities"* is not warranted. Juan Martinez Alier [50] , for instance shows that the bulk of consumers are automatically excluded from having an impact upon the prices of commodities, as these consumers are future generations who have not been born yet. The assumptions behind future discounting, which assume that future goods will be cheaper than present goods, has been criticised by Fred Pearce[51] and by the recent Stern Report (although the Stern report itself does employ discounting and has been criticised by ecological economists).[52] Concerning these externalities, Paul Hawken argues that the only reason why goods produced unsustainably are usually cheaper than goods produced sustainably is due to a hidden subsidy, paid by the non-monetized human environment, community or future generations[53] . These arguments are developed further by Hawken, Amory and Hunter Lovins in "Natural Capitalism: Creating the Next Industrial Revolution"[54] .

See also

- Agroecology
- → Deep ecology
- Eco socialism
- Ecofeminism
- Ecology of contexts
- Embodied energy
- Embodied water
- Energy economics
- Environmental economics
- → Environmental ethics
- Green accounting
- *The Green Economist* (news journal)
- Human development theory
- Human ecology
- Inclusive Democracy
- Industrial ecology
- Natural capital
- Natural resource economics
- Sustainability
- Spaceship Earth
- Thermoeconomics
- Energy Accounting
- Economics and energy
- Exergy
- List of Green topics
- Energy quality
- Limits to growth
- Steady state economics

Further reading

- Common, M. and Stagl, S. 2005. *Ecological Economics: An Introduction*. New York: Cambridge University Press.
- Costanza, R., Stern, D. I., He, L., Ma, C. (2004). Influential publications in ecological economics: a citation analysis. Ecological Economics 50(3-4): 261-292. - http://econpapers.repec.org/article/eeeecolec/ v_3A50_3Ay_3A2004_3Ai_3A3-4_3Ap_3A261-292.htm
- Daly, H. and Townsend, K. (eds.) 1993. *Valuing The Earth: Economics, Ecology, Ethics*. Cambridge, Mass.; London, England: MIT Press.
- Georgescu-Roegen, N. 1975. Energy and economic myths. *Southern Economic Journal* 41: 347-381.
- Krishnan R, Harris JM, Goodwin NR. (1995). *A Survey of Ecological Economics*. Island Press. ISBN 1559634111, 9781559634113.
- Martinez-Alier, J. (1990) Ecological Economics: Energy, Environment and Society. Oxford, England: Basil Blackwell.
- Martinez-Alier, J., Ropke, I. eds. (2008). Recent Developments in Ecological Economics, 2 vols., E. Elgar, Cheltenham, UK.
- Røpke, I. (2004) The early history of modern ecological economics. Ecological Economics 50(3-4): 293-314.

- Røpke, I. (2005) Trends in the development of ecological economics from the late 1980s to the early 2000s. Ecological Economics 55(2): 262-290.
- Spash, C. L. (1999) The development of environmental thinking in economics. Environmental Values 8(4): 413-435.
- Stern, D. I. (1997) Limits to substitution and irreversibility in production and consumption: A neoclassical interpretation of ecological economics. Ecological Economics 21(3): 197-215. - http://econpapers.repec.org/article/eeeecolec/v_3A21_3Ay_3A1997_3Ai_3A3_3Ap_3A197-215.htm
- Tacconi, L. (2000) Biodiversity and Ecological Economics: Participation, Values, and Resource Management. London, UK: Earthscan Publications.
- Vatn, A. (2005) Institutions and the Environment. Cheltenham: Edward Elgar.

External links

- The International Society for Ecological Economics (ISEE) - http://www.ecoeco.org/
  - The academic journal, *Ecological Economics* - http://www.elsevier.com/locate/ecolecon
  - Ecological Economics Encyclopedia - http://www.ecoeco.org/education_encyclopedia.php
  - The US Society of Ecological Economics - http://www.ussee.org/
  - The Brazilian Society for Ecological Economics (ECOECO) - http://www.ecoeco.org.br/

Schools and institute:

- The Gund Institute of Ecological Economics - http://www.uvm.edu/giee
- Ecological Economics at Rensselaer Polytechnic Institute - http://www.economics.rpi.edu/ecological.html
- The Beijer International Institute for Ecological Economics - http://www.beijer.kva.se/
- The Green Economics Institute- http://www.greeneconomics.org.uk/publishes the

International Journal of Green Economics http://www.inderscience.com/ijge/

Environmental data:

- EarthTrends World Resources Institute - http://earthtrends.wri.org/index.php
- Eco-Economy Indicators: http://www.earth-policy.org/Indicators/index.htm
- NOAA Economics of Ecosystems Data & Products – http://www.economics.noaa.gov/?goal=ecosystems

Miscellaneous:

- Green Economics website- http://greeneconomics.org/
- Gaian Economics website - http://www.gaianeconomics.org/
- Sustainable Prosperity - http://sustainableprosperity.ca/
- World Resources Forum - http://www.worldresourcesforum.org
- An ecological economics article about reconciling economics and its supporting ecosystem - http://www.fs.fed.us/eco/s21pre.htm
- "Economics in a Full World", by Herman E. Daly - http://sef.umd.edu/files/ScientificAmerican_Daly_05.pdf
- Steve Charnovitz, "Living in an Ecolonomy: Environmental Cooperation and the GATT," Kennedy School of Government, April 1994.

References

[1] Ott, K. (2003). "The Case for Strong Sustainability." (http://umwethik.botanik.uni-greifswald.de/booklet/8_strong_sustainability.pdf) In: Ott, K. & P. Thapa (eds.) (2003).*Greifswald's Environmental Ethics*. Greifswald: Steinbecker Verlag Ulrich Rose. ISBN 3931483320. Retrieved on: 2009-02-16.

[2] Anastasios Xepapadeas (2008). "ecological economics," *The New Palgrave Dictionary of Economics*, 2nd Edition. Abstract. (http://pde-aux1.pde.pm.semcs.net/article?id=pde2008_E000221&q=bioeconomics&topicid=&result_number=4)

[3] Jeroen C.J.M., Bergh, van den (2000). Ecological Economics: Themes, Approaches, and Differences with Environmental Economics. Tinbergen Institute Discussion Paper TI 2000-080/3. pp. 1-25. (http://www.tinbergen.nl/discussionpapers/00080.pdf) Accessed: 2009-07-02.

[4] Illge L, Schwarze R. (2006). A Matter of Opinion: How Ecological and Neoclassical Environmental Economists Think about Sustainability and Economics (http://www.diw.de/deutsch/produkte/publikationen/diskussionspapiere/docs/papers/dp619.pdf). German Institute for Economic Research.

[5] Paehlke R. (1995). Conservation and Environmentalism: An Encyclopedia, p. 315 (http://books.google.com/books?id=9WUqqgfrBHQC&printsec=frontcover#PRA2-PA315,M1). Taylor & Francis.

[6] Scott Cato, M. (2009). *Green Economics*. Earthscan, London. ISBN 9781844075713.

[7] Malte Faber. (2008). How to be an ecological economist. *Ecological Economics* 66(1):1-7. Preprint (http://ideas.repec.org/p/awi/wpaper/0454.html).

[8] Peter Victor. (2008). Book Review: Frontiers in Ecological Economic Theory and Application. *Ecological Economics* 66(2-3).

[9] Mattson L. (1975). Book Review: *Positional Analysis for Decision-Making and Planning* by Peter Soderbaum. *The Swedish Journal of Economics*.

[10] Soderbaum, P. 2008. *Understanding Sustainability Economics*. Earthscan, London. ISBN 9781844076277. pp.109-110, 113-117.

[11] Harris J. (2006). *Environmental and Natural Resource Economics: A Contemporary Approach*. Houghton Mifflin Company.

[12] Costanza R et al. (1998). " The value of the world's ecosystem services and natural capital1 (http://linkinghub.elsevier.com/retrieve/pii/S0921800998000202)". *Ecological Economics* 25 (1): 3–15. doi: 10.1016/S0921-8009(98)00020-2 (http://dx.doi.org/10.1016/S0921-8009(98)00020-2). .

[13] Knapp G, Roheim CA and Anderson JL (2007) *The Great Salmon Run: Competition Between Wild And Farmed Salmon* (http://search.worldwildlife.org/cs.html?url=http://www.worldwildlife.org/what/globalmarkets/wildlifetrade/WWFBinaryitem4985.pdf&qt=The+Great+Salmon+Run&col=&n=4) World Wildlife Fund. ISBN 0-89164-175-0

[14] Washington Post. Salmon Farming May Doom Wild Populations, Study Says (http://www.washingtonpost.com/wp-dyn/content/article/2007/12/13/AR2007121301190.html).

[15] Soil Association. UK Organic Group Exposes Myth that Genetically Engineered Crops Have Higher Yields (http://www.organicconsumers.org/articles/article_11493.cfm). Organic Consumers Association.

[16] Soderbaum P. (2004). Politics and Ideology in Ecological Economics (http://www.ecoeco.org/pdf/politics_ideology.pdf). Internet Encyclopaedia of Ecological Economics.

[17] Costanza R. (1989). What is ecological economics? *Ecological Economics* 1:1-7. Free full text (http://www.uvm.edu/giee/publications/Costanza_EE_1989.pdf).

[18] http://www.ecoeco.org/pdf/costanza.pdf

[19] http://unjobs.org/authors/juan-martinez-alier

[20] Costanza R. (2003). Early History of Ecological Economics and ISEE (http://www.ecoeco.org/pdf/costanza.pdf). Internet Encyclopaedia of Ecological Economics.

[21] Kapp, K. W. (1950) The Social Costs of Private Enterprise. New York: Shocken.

[22] Polanyi, K. (1944) The Great Transformation. New York/Toronto: Rinehart & Company Inc.

[23] Schumacher, E.F. 1973. *Small Is Beautiful: A Study of Economics as if People Mattered*. London: Blond and Briggs.

[24] Daly, H. 1991. *Steady-State Economics* (2nd ed.). Washington, D.C.: Island Press.

[25] http://www.nytimes.com/2009/04/12/opinion/12zencey.html?pagewanted=1&_r=2&ref=opinion Eric Zencey, a professor of historical and political studies at Empire State College. A version of this article appeared in print on April 12, 2009, on page WK9 of the New York edition.

[26] Georgescu-Roegen, N. 1971. *The Entropy Law and the Economic Process*. Cambridge, Mass.: Harvard University Press.

[27] Røpke, I. (2004) The early history of modern ecological economics. Ecological Economics 50(3-4): 293-314. Røpke, I. (2005) Trends in the development of ecological economics from the late 1980s to the early 2000s. Ecological Economics 55(2): 262-290.

[28] Spash, C. L. (1999) The development of environmental thinking in economics. Environmental Values 8(4): 413-435.

[29] Costanza, R., Stern, D. I., He, L., Ma, C. (2004). Influential publications in ecological economics: a citation analysis. Ecological Economics 50(3-4): 261-292.

[30] http://en.wikipedia.org/wiki/Template:Thermodynamic

[31] Norgaard, R. B. (1994) Development Betrayed: The End of Progress and a Coevolutionary Revisioning of the Future. London: Routledge

[32] Daily, G.C. 1997. *Nature's Services: Societal Dependence on Natural Ecosystems*. Washington, D.C.: Island Press.

[33] Millennium Ecosystem Assessment. 2005. *Ecosystems and Human Well-Being: Biodiversity Synthesis*. Washington, D.C.: World Resources Institute.

[34] McCauley, D. J. (2006) Selling out on nature. Nature 443(7): 27-28

[35] Daly, H. and Farley, J. 2004. *Ecological Economics: Principles and Applications*. Washington: Island Press.

[36] Peter A. Corning 1 *, Stephen J. Kline. (2000). Thermodynamics, information and life revisited, Part II: Thermoeconomics and Control information (http://www3.interscience.wiley.com/cgi-bin/abstract/71007254/ABSTRACT) Systems Research and Behavioral Science, Apr. 07, Volume 15, Issue 6 , Pages 453 – 482

[37] Corning, P. (2002). " Thermoeconomics – Beyond the Second Law (http://www.complexsystems.org/abstracts/thermoec.html)" – source: www.complexsystems.org

[38] http://exergy.se/goran/thesis/ *Exergy - a useful concept* by Göran Wall

[39] http://telstar.ote.cmu.edu/environ/m3/s3/05account.shtml Environmental Decision making, Science and Technology

[40] Stabile, Donald R. "Veblen and the Political Economy of the Engineer: the radical thinker and engineering leaders came to technocratic ideas at the same time," *American Journal of Economics and Sociology (45:1) 1986, 43-44.*

[41] Costanza, R., d'Arge, R., de Groot, R., Farber, S., Grasso, M., Hannon, B., Naeem, S., Limburg, K., Paruelo, J., O'Neill, R.V., Raskin, R., Sutton, P., and van den Belt, M. 1997. The value of the world's ecosystem services and natural capital. *Nature* 387: 253-260.

[42] Norgaard, R.B. and Bode, C. 1998. *Next, the value of God, and other reactions. Ecological Economics* 25: 37-39.

[43] http://www.socialedge.org/features/opportunities/archive/2008/02/23/weblogentry.2008-02-20.6191038944

[44] http://www.sosyasuni.org/en/News/Ecuadors-Oil-Change-An-Exporters-Historic-Proposal.html

[45] http://www.multinationalmonitor.org/mm2007/092007/koenig.html

[46] http://www.cnn.com/2007/BUSINESS/12/10/ecuador.oil.ap/

[47] http://abcnews.go.com/International/wireStory?id=3980994

[48] http://www.panna.org/jt/agAssessment

[49] http://www.telegraph.co.uk/opinion/main.jhtml?xml=/opinion/2008/04/17/do1702.xml

[50] Jose Maria Figueres Olson (Foreword), Robert Costanza (Editor), Olman Segura (Editor), Juan Martinez-Alier (Editor), Juan Martinez Alier (Author) (1996) "Getting Down to Earth: Practical Applications Of Ecological Economics" (Intl Society for Ecological Economics) (Island Press)

[51] Pearce, Fred "Blueprint for a Greener Economy"

[52] Spash, C. L. (2007) The economics of climate change impacts à la Stern: Novel and nuanced or rhetorically restricted? Ecological Economics 63(4): 706-713

[53] Hawken, Paul (1994) "The Ecology of Commerce" (Collins)

[54] Hawken, Paul; Amory and Hunter Lovins (2000) "Natural Capitalism: Creating the Next Industrial Revolution" (Back Bay Books)

Ecology

The science of **ecology** includes everything from global processes (above), the study of various marine and terrestrial habitats (middle) to individual interspecific interactions like predation and pollination (below).

Ecology (from Greek: οἶκος, "house" ; -λογία, "study of") is the interdisciplinary scientific study of the interactions between organisms and their environment.[1] Ecology is also the study of ecosystems. Ecosystems describe the web or network of relations among organisms at different scales of organization. Since ecology refers to any form of biodiversity, ecologists can conduct research on the smallest bacteria to the global flux of atmospheric gases that are regulated by photosynthesis and respiration as organisms breath in and out of the biosphere. Ecology is a recent discipline that emerged from the natural sciences in the late 19th century. Ecology is not synonymous with environment, → environmentalism, or environmental science.[1] [2] [3]

Like many of the natural sciences, a conceptual understanding of ecology is found in the broader details of study, including:

- life processes explaining adaptations
- distribution and abundance of organisms
- the movement of materials and energy through living communities
- the successional development of ecosystems, and
- the abundance and distribution of biodiversity in context of the environment.[1] [2] [3]

Ecology is distinguished from natural history, which deals primarily with the descriptive study of organisms. It is a sub-discpline of biology, which is the study of life.

There are many practical applications of ecology in conservation biology, wetland management, natural resource management (agriculture, forestry , fisheries), city planning (urban ecology), community health, → economics, basic & applied science and it provides a conceptual framework for understanding and researching human social interaction (human ecology).[4] [5] [6] [7]

Levels of organization and study

Ecology is challenged by a constant analytical problem of how to deal with different scales of pattern in space and time. Ecological processes can take decades and even hundreds of years to mature and cover broad geographic areas. Ecologists study ecosystems by sampling a certain number of individuals that are representative of a population. Long-term ecological studies, such as sites managed by the Long Term Ecological Network [8] including the Hubbard Brook study in operation since 1960 [9], provide important ecological track records; the longest experiment in existence is the 'Park Grass Exeriment' that started in 1856[10] . Most studies, however, cover only a fraction of the life-span in the development of an ecosystem, such as the different seral stages leading up to an old-growth forest. Ecology is also complicated by the fact that small scale patterns do not necessarily explain large scale phenomena, otherwise captured in the expression 'the sum is greater than the parts'.[11] These emergent phenomena operate at different environmental scales of influence, ranging from molecular to galactic spheres, and require different sets of scientific explanation.[12] [13]

To simplify and place the study of ecology into a manageable framework of understanding, the biological world is conceptually organized as a nested hierarchy of individuality, ranging in scale from genes, to cells, to tissues, to organs, to organisms, to species and up to the level of the biosphere.[14] Ecosystems are primarily researched at (but not restricted to) three key levels of organization, including (1) organisms, (2) populations, and (3) communities. Ecosystems consist of communities containing different species of organisms. Communities consist of organisms living in different populations. [15] [16] Ecosystem diversity is a part of biodiversity.

Ecosystems regenerate after a disturbance, such as fire, and form mosaics of different age groups structured across a landscape. This image shows different seral stages in forested ecosystems starting from pioneers colonizing a disturbed site and maturing through time in successional stages leading to old-growth forests.

Biodiversity includes all the varieties and processes of life, including organisms and their genetic differences that are evolutionarily classified into hierarchical, branching and coalescing dimensions.[17] [18] [19]

Ecological niche

We are not here concerned with an imaginary ecology based upon a hypothetical environment inhabited by fancied organisms evolved in some vaguely conceived system of life [...] Instead of dealing with imaginary situations, we are confronted by the ecology of the Earth as we know it, populated by organisms that have evolved here from the basis furnished principally by water, carbon dioxide, and their elements, together with nitrogen.

—Allee et al.[2] :73

The ecological niche is a central concept in ecology. There are many definitions of the niche dating back to 1917[20], but George Evelyn Hutchinson made conceptual advances on the concept in 1957[21] [22] and introduced the most widely accepted definition:

> "The niche is the set of biotic and abiotic conditions in which a species is able to persist and maintain stable population sizes."[20] :519

There are two differentiated kinds of ecological niche known as the *fundamental* and the *realized* niche. The fundamental niche describes the abiotic conditions under which a species is able to persist. The realized niche is the set of conditions under which a species persists in the context of other resource competitors or predators.[22] [20] [15] Organisms fit into a particular ecological niche according to their functional traits. A trait is a measurable property of an individual that strongly influences its performance.[23] Species become specialized within their niche and competitively exclude other species from living in the same geographic area if they fit into the same ecological niche. This is called the competitive exclusion principle.[24] Equally important to the concept of niche is habitat. The habitat describes the environment over which a species is known to occur and the type of community that is formed as a result.[25] For example, habitat might refer to an aquatic versus terrestrial environment that can be further categorized as montane or alpine.

Some of the biodiversity of a coral reef. Corals modify their environment by laying down calcium carbonate skeletons that provide growth foundations for future generations and builds structural habitats for many other species.

Organisms are subject to environmental pressures, but they are also modifiers of their habitats. The regulatory feedback relationship between organisms and their environment can significantly modify conditions from a local scale (e.g., a pond) to global scale (e.g., Gaia) and they can also modify conditions over time even after an organism has passed away, such as the remnants of an old beaver dam or silica skeleton deposits from marine organisms.[26] This process of ecosystem engineering has also been called niche construction. Ecosystem engineers are defined as:

> "...organisms that directly or indirectly modulate the availability of resources to other species, by causing physical state changes in biotic or abiotic materials. In so doing they modify, maintain and create habitats." [27] :373

Although it has long been understood that organisms modify their environment, the ecological engineering concept has stimulated a new appreciation for the degree of modification and the influence organisms have on the ecosystem and evolutionary process.[28] [29] The niche construction concept highlights a previously under appreciated feedback mechanism of natural selection imparting forces on the abiotic niche.

"For example, many ant and termite species regulate temperature by plugging nest entrances at night or in the cold, by adjusting the height or shape of their mounds to optimize the intake of the sun's rays, or by carrying their brood around their nest to the place with the optimal temperature and humidity for the brood's development."[29] :10242

Population ecology

The first journal publication of the Society of Population Ecology, titled Population Ecology (originally called Researches on Population Ecology), was released in 1952.[30] Population ecology is concerned with the study of groups of organisms that live together in time and space. One of the first laws of population ecology is the Thomas Malthus' exponential law of population growth.[31] This law states that:

> "...a population will grow (or decline) exponentially as long as the environment experienced by all individuals in the population remains constant."[31] :18

> At its most elementary level, interspesific competition involves two species utilizing a similar resource. It rapidly gets more complicated, but stripping the phenomenon of all its complications, this is the basic principal: two comnsumers consuming the same resource.[32] :222

This simplified premise in population ecology provides the basis for formulating predictive theories and tests that follow. Simplified models in population ecology usually start with four key variables including death, birth, immigration, and emigration. The ecology of populations are simplified in the mathematical models that calculate changes in population demographics and evolution under the assumption (or null hypothesis) of no external influence. Some models can become more mathematically complex where "...several competing hypotheses are simultaneously confronted with the data."[33] For example, in a closed system where immigration and emigration does not take place, the per capita rates of change in a population can be mathematically described as:

$$dN/dT = B - D = bN - dN = (b - d)N = rN,$$

where N is the total number of individuals in the population, B is the number of births, D is the number of deaths, b and d are the per capita rates of birth and death respectively, and r is the per capita rate of population change. In simple terms, this formula can be understood as the rate of change in the population (dN/dT) is equal to births minus deaths (B - D).[31] [32]

Using these techniques, Malthus' population principal of growth was later transformed into a mathematical model known as the logistic equation:

$$dN/dT = aN(1 - N/K),$$

where N is the biomass density, a is the maximum per-capita rate of change, and K is the carrying capacity of the population. The formula can be read as follows, the rate of change in the population (dN/dT) is equal to growth (aN) that is limited by carrying capacity ($1-N/K$). From these basic mathematical principals the discipline of population ecology expands into a field of investigation that queries the demographics of real populations and tests these results against those of various statistical models. Beyond these, the field of population ecology often uses data on life history and matrix algebra to develop projection matrices on fecundity and survivorship. This kind of information can be used for managing wildlife stocks and harvest quotas [34] [32]

Terms used to describe natural groupings of individuals in ecological studies[35]

Term	Definition
Species population	All individuals of a species.
Metapopulation	A set of spatially disjunct populations, among which there is some immigration.
Population	A group of conspecific individuals that is demographically, genetically, or spatially disjunct from other groups of individuals.
Aggregation	A spatially clustered group of individuals.
Deme	A group of individuals more genetically similar to each other than to other individuals, usually with some degree of spatial isolation as well.

| Local population | A group of individuals within an investigator-delimited area smaller than the geographic range of the species and often within a population (as defined above). A local population could be a disjunct population as well. |
| Subpopulation | An arbitrary spatially-delimited subset of individuals from within a population (as defined above). |

These mathematical models introduce two important variables that are commonly invoked in population ecology, namely r (intrinsic rate of natural increase in population size, density independent) and K (carrying capacity of a population, density dependent).[15] These two variables where used in development of the concept of r and K selection. An r-selected species (e.g., many kinds of insects, such as aphids[36]) is one that has high rates of fecundity, low levels of parental investment in the young, and high rates of mortality before individuals reach maturity. In r-selected species evolution favors productivity. In contrast, a K-selected species (such as humans) has low rates of fecundity, high levels of parental investment in the young, and low rates of mortality as individuals develop toward maturity. Evolution in K-selected species favors efficiency in the conversion of resources into fewer offspring.[37] [38]

Populations are also studied and conceptualized through the metapopulation concept. The metapopulation concept was introduced in 1969[39] :

> "as a population of populations which go extinct locally and recolonize."[40] :105

Metapopulation ecology is a simplified model of the landscape into patches of varying levels of quality.[41] These patches in the landscape are either occupied or they are not. Migrants moving among the patches are structured into metapopulations either as sources or sinks. Source patches are productive sites that generate a seasonal supply of migrants to other patch locations. Sink pathes are unproductive sites that only receive migrants. In metapopulation terminology there are emigrants (individuals that leave a patch) and immigrants (individuals that move into a patch). Metapopulation models examine patch dynamics over time, investigating questions about spatial ecology. An important concept in metapopulation ecology is the rescue effect, where small patches of lower quality (i.e., sinks) are maintained by a seasonal influx of new immigrants. Natural seasonal dynamics of ecology means that metapopulation structure evolves from year to year, where some patches are sinks during dry years, for example, and become sources when conditions are more favorable. Metapopulation ecologists utilize a mixture of computer models and ecological field studies to understand the spatial distributions of organisms and the ecological processes that help to explain the metapopulation structure.[42]

Community ecology

Community ecology examines how interactions among species and their environment affect the abundance, distribution and diversity of species within communities.

—Johnson & Stinchcomb[43] :250

Ecosystems are most generally studied at the local or effective community scale, such as measuring primary production in a wetland in relation to decomposition and consumption rates[44] or the analysis of predator-prey dynamics affecting amphibian biomass[45] . The vast majority of research into community ecology examines population dynamics of pairs of species to understand how entire communities function. Two conceptual models that have been used in understanding community ecology include food webs and trophic levels.[46] [47]

Food webs

A schematic illustration of a salamander food-web in a pond (left) and an energy flow diagram from the Silver Springs community study (right).[48]

Food webs are a type of concept map that are used to understand and map out real pathways in the series of ecological events usually starting with solar energy being photosynthesized in plants. Plants grow and accumulate nutrients that are in turn eaten by grazing herbivores and step by step the lines are drawn and until the web of life is illustrated.[49] [50]

The first person to fully elaborate and place the concept of food chains into a scientific framework was Charles Elton in his classical book 'Animal Ecology'.[51] Elton[51] defined ecological relations using concepts of food-chains, food-cycles, food-size, and described numerical relations among different functional groups and their relative abundance. Elton's term 'food-cycle' was replaced by 'food-web' in a subsequent ecological text[52]. Elton's book broke conceptual ground by illustrating complex ecological relations through simpler food-web diagrams.[49] Food-webs are an effective way to conceptually illustrate and teach about the interactive links among species in a community.[53] [54]

A simplified illustration of a terrestrial and an ocean food-chain.

There are different dimensions in ecological communities that can be used to create more complicated food-webs, including: species composition (type of species), richness (number of species), biomass (the dry weight of plants and animals), productivity (rates of conversion of energy and nutrients into growth), and stability (food-webs over time). A food-web diagram illustrating species composition shows how a change in one single species can directly and indirectly influence many others. Microcosm studies are used to simplify food-web research into semi-isolated units such as small springs, decaying logs and in laboratory experiments using organisms that reproduce quickly, such as daphnia feeding on algae grown under controled environments in jars of water.[55] [56] Principals gleaned from these food-web microcosm studies are used to extrapolate smaller dynamic concepts to larger systems.[56] Food-chain length is an important parameter in describing larger food-web dynamics and is defined as:

"The number of transfers of energy or nutrients from the base to the top of a food web..."[57] :269

There are different ways of calculating food-chain length depending on what parameters of the food-web dynamic are being considered: connectance, energy, or interaction.[57] Hence, in a simple predator-prey example a deer is one step removed from the plants it eats (chain length = 1) and a wolf that eats the deer is two steps removed (chain length = 2). The relative amount or strength of influence that these parameters have on the food-web are used to address questions about:

- the identity or existence of a few dominant species (called strong interactors or keystone species)
- the total number of species and food-chain length (including many weak interactors) and
- how community structure, function and stability is determined.[56]

Trophic dynamics

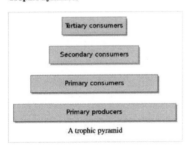

A trophic pyramid

Links in food-webs relate of primary importance to feeding relations or trophism (The Greek root of the word troph, τροφή, trophē, means food or feeding). When the relative abundance of each functional feeding group is stacked into their respective trophic levels they naturally sort into a 'pyramid of numbers'.[51] Functional groups are broadly categorized as autotrophs (e.g., plants), heterotrophs (e.g., deer, wolves), and detrivores (e.g., bacteria, fungi). Some organisms are omnivores, meaning they eat both plant and animal tissues and don't fit neatly into a category. However, it has been suggested that omnivores have a greater functional ecosystem influence as predators because relative to herbivores they are comparatively inefficient at grazing.[58]

Functional groups are usually depicted in hierarchical schemes with three or more trophic levels including primary producers (autotrophs) and levels of heterotrophic consumers including the herbivores (primary consumers), predators (secondary consumers), predators that eat predators (tertiary consumers), and ultimately ending at the detrivores in the soil ecosystems.[59] The pyramidal arrangement of trophic levels is a consistent feature across ecosystems with the primary producers having the larger base and consumer densities and amounts of energy decreasing as species become further removed from the photosynthetic source of production.[60] The size of each level in the pyramid generally represents biomass, which is often measured as the dry weight of an organism.[59] Trophic levels and food webs can be used to statistically model or mathematically calculate parameters such as those used in other kinds of network analysis, including graph theory.[61]

List of ecological functional groups, definitions and examples

Functional Group	Definition and Examples
Producers or *Autotrophs*	Usually plants or cyanobacteria that are capable of photosynthesis but could be other organisms such as the bacteria near ocean vents that are capable of chemosynthesis.
Consumers or *Heterotrophs*	Animals, which can be primary consumers (herbivorous), or secondary or tertiary consumers (carnivorous and omnivores).
Decomposers or *Detritivores*	Bacteria, fungi, and insects which degrade organic matter of all types and restore nutrients to the environment. The producers will then consume the nutrients, completing the cycle.

Autotrophs have the highest global proportion of biomass, rivaled by microbes (prokaryotes - decomposers) and then heterotrophs.[62] [63] Functional trophic groups sort out hierarchically into pyramidic trophic levels because it requires specialized adaptations to become a photosynthesizer or a predator, but rarely are their organisms having a skillful combination of both functional abilities. This explains why functional adaptations to trophism (feeding) organizes different species into emergent functional groups.[58]

Trophic levels are part of the holistic or complex systems view of ecosystems.[64] [65] Each trophic level contains different species that share common ecological functions. Different species, such as ferns and lillys, are grouped very differently from an evolutionary view of their ecology, but functionally they both photosynthesize the sun's energy and classified as autotrophs. Grouping these functionally similar species into a trophic system gives a macroscopic image of the larger functional design. Trophic levels are abstractions of the system, but they explain real phenomena.[66]

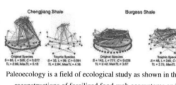

Paleoecology is a field of ecological study as shown in these reconstructions of fossilized food-web ecosystems and trophic levels sorted vertically. Primary producers are at the base as red spheres, predator's are at the top as yellow spheres, and the lines represent feeding links. Original food-webs (on the left) are simplified on the right panels by aggregating groups that feed on the same foods into trophic species.[67]

Food-web links point to direct trophic relationships among species, but there are also indirect effects that can alter the abundance, distribution, or biomass in the trophic levels. For example, predators eating herbivores indirectly influence the control and regulation of primary production in plants. Although the predators do not eat the plants directly, they regulate the population of herbivores that are directly linked to plant trophism. The net effect of direct and indirect relations is called trophic cascades. Trophic cascades are separated into species-level cascades, where only a subset of the food-web dynamic is impacted by a change in population numbers, and community-level cascades, where a change in population numbers has a dramatic effect on the entire food-web, such as the distribution of plant biomass.[68]

The keystone species concept is closely aligned to species-level cascades, where a single species occupies a particularly strong node in the food-web and its removal results in the collapse of the food-web structure and extinction of other species. Sea otters (*Enhydra lutris*) are the classical example of a keystone species because they limit the density of urchins that feed on kelp. If sea otters are removed from the system, the urchins graze until the kelp beds disappear and this has a dramatic effect on community structure.[69] Hunting of sea otters, for example, is thought to have indirectly lead to the extinction of the Steller's Sea Cow (*Hydrodamalis gigas*).[70] While the keystone species concept has been used extensively as a conservation tool, it has been criticized for being poorly defined. Different ecosystems express different complexities and so it is unclear how applicable and general the keystone species model can be applied. To better understand the keystone species and trophic cascade models, ecologists conduct removal experiments to measure the relative impact, strength and influence of interaction among different species on community dynamics.[68] [69]

Biosphere

The largest scale of ecological organization is the total sum of every ecosystem on the planet and the atmosphere it regulates, which is called the biosphere. Ecological relations regulate the flux of energy, nutrients, and climate all the way up to the planetary scale. For example, the dynamic history of the planetary CO_2 and O_2 composition of the atmosphere has been largely regulated by the biogenic flux of gases coming from respiration and photosynthesis with levels fluctuating in time and in relation to the ecology and evolution of plants and animals.[71] When sub-component parts, such as the full variety of ecosystems diversifying the planet, are organized into a whole there are oftentimes identifiable properties or characteristics that describe the nature of the system under investigation. Ecological theory has been used to explain self emergent regulatory phenomena at the planetary scale. This is known as the Gaia hypothesis[13] . The Gaia hypothesis is an example of holism in ecology because it tests for principals relating to an evolving and self regulating planetary ecosystem that requires different explanations than those governing ecosystems at a smaller scale.[72]

Ecology and evolution

Ecology and evolution are considered sister disciplines. Ecology and evolution are academic branches of the life sciences. Natural selection, life history, development, adaptation, populations, and inheritance all play an prominent conceptual roles in ecological as well as evolutionary theory. Both disciplines also employ genetics in their investigations. For example, morphological, behavioural and/or genetic traits can be mapped onto evolutionary trees to study principals of inheritance that relates back to the ecology of adaptations. Ecology and evolution are scientifically connected because they both study hierarchies, networks, relations, and kinship among genes, cells, individuals, communities, species, and the biosphere.[73] The two disciplines often appear together, such as in the title of the journal *Trends in Ecology and Evolution*.[74] There is no sharp dichotomous boundary that separates ecology from evolution and differ more in their areas of applied focus. Both disciplines discover and explain emergent and unique properties and processes operating across different spatial or temporal scales of organization.[75] [13] [76] While the boundary between ecology and evolution is not always clear, it is understood that ecologists study the abiotic and biotic factors that influence the evolutionary process.[2] [59]

Behavioral ecology

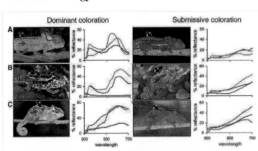

Social display and color variation in differently adapted species of chameleons (*Bradypodion* spp.). Chameleons can change their skin color to match their background as a behavioral defense mechanism and can also use color to signal to other members of their species, such as dominant (left) versus submissive (right) patterns shown in the three species (A-C) above.[77]

Behavioural ecology is the field of study concerned with ethology and its implications to broader ecological theory. Adaptation is the central unifying concept in behavioral ecology.[78] Behaviors can be recorded as traits and inherited in much the same way that eye and hair color can. As such, behaviors are subject to the forces of natural selection.[79] Hence, behaviors can be adaptive in nature, meaning that they evolved and serve a functional utility such as enhancing ones opportunity to successfully reproduce and increase fitness.[80] Fitness is measured in terms of reproductive success. An animal with behaviors that afford it some degree of leverage in the struggle for existence such that it survives to pass on its heritable traits to its offspring is considered fit if the adaptation succeeds and propagates more of its kind in subsequent generations. A measure of fitness is the numerical differential and representation in frequency of a trait over subsequent generations.[79]

Predator-prey interactions are a fundamental and introductory concept in food-web studies as well as behavioural ecology.[81] Prey species can exhibit different kinds of behavioural adaptations to predators, such as avoid, flee or defend. Many prey species are faced with multiple predators that differ in the degree of danger posed. To be adapted to their environment and predatory threats, organisms must balance their energy budgets as they invest in different aspects of their life history, such as growth and feeding, mating, socializing, or modifying their habitat. Hypotheses posited in behavioural ecology are generally based on adaptive principals of conservation or efficiency.[82] [1] [2] [3] For example,

"The threat-sensitive predator avoidance hypothesis predicts that prey should assess the degree of threat posed by different predators and match their behavior according to current levels of risk."[83]

"The optimal flight initiation distance occurs where expected postencounter fitness is maximized, which depends on the prey's initial fitness, benefits obtainable by not fleeing, energetic escape costs, and expected fitness loss due to predation risk."[84]

The behaviour of long-toed salamanders (*Ambystoma macrodactylum*) present another example in this context. When threatened, the long-toed salamander will defend itself by waving its tail and secreting a white milky fluid.[85] [86] The excreted fluid is distasteful, toxic and adhesive, but it is also used for nutrient and energy storage during hibernation. Hence, salamanders subjected to frequent predatory attack will be energetically compromised as they use up their energy stores.[87] [88] Some species are also able to avoid predators altogether, such as small velvet gecko's (*Oedura lesueurii*). This species is specially adapted to smell the body chemicals of snakes that linger after they pass through an area, even though snakes rarely pose a significant danger.[83]

Many creatures also form mutually beneficial relationships called mutualisms.[89] Approximately 60% of all plants, for example, form a symbiotic relationship with arbuscular mycorrhizal fungi. The plants and fungi exchange carbohydrates for mineral nutrients.[90] The range of relationships that can form in ecology can be simplfied down to the *host* and the *associate*. A host is any entity that harbors another that is called the associate. A genome is the host for the associate genes in the same way that parasites are associates on host organisms.[91] The are different types of ecological interactions that can reciprocally develop from the host and associate relationship. If the relationship is mutually beneficial it is called mutualism, one species benefits and the other suffers, such as in parasitism or predation. Competition is reciprocal antagonism among species. Symbiosis is a form of mutualism where the organisms are physically connected, whereas there are mutualisms that form out of indirect relations.[92]

Leafhoppers are protected by an army of meat ants in a symbiotic relationship.

	Species 1		
	+	-	0
Species 2 +	Mutualism		
-	Predation/Parasitism	Competition	
0	Commensalism	Amensalism	Neutralism

A grid of type of relations that can evolve among species, where antagonistic relations are denoted by a minus sign, beneficial relations are denoted by a plus sign, and neutral relations are denoted by the zero's.[93]

There are many examples in all corners of life of interspecific mutualistic relations. Famous ecological study systems where mutualism occurs include, fungus-growing ants with agricultural like societies, bacteria living in the guts of insects and other organisms, the fig wasp and yucca moth pollination complex, and corals with their photosynthetic algae.[94] [95]

Intraspecific behaviours are most notable in the social insects, slime moulds, social spiders, human society, and naked mole rats where eusocialism has evolved. Social behaviours include reciprocally beneficial behaviours among kin and nest mates.[96] [79] [97] Social behaviours evolve from kin and group selection. Kin selection explains altruism through genetic relationships, whereby an altruistic behaviour leading to death is rewarded by the survival of genetic copies distributed among surviving relatives. The social insects, including ants, bees and wasps are most famously studied for this type of relationship because the male drones are clones that share the same genetic make-up as every other male in the colony.[79] In contrast, group selectionists find examples of altruism among non-genetic relatives and explain this through selection acting on the group, whereby it becomes selectively advantageous for a group if its members express altruistic behaviours to one another. Groups that are predominantely altruists beat groups that are predominantely selfish.[79] [98]

A often quoted hypothesis in behavioural ecology is known as Lack's brood reduction hypothesis, which posits an evolutionary and ecological explanation as to why birds often lay a series of eggs with an asynchronous delay such that the young are of mixed age and weights. According to Lack, this brood-reducation behaviour is a sort of ecological insurance that allows some birds to survive in poor years and all birds to survive when food is plentiful.[99] [100]

"The clutch size of each species of bird is characteristic, and in general seems adapted to correspond with the largest number of young which can be successfully raised. Probably there is a small hereditary variation, which is explicable through rather larger clutches being favoured in some years and rather smaller clutches in other years."[101] :333

Elaborate sexual displays and posturing are often encountered in the behavioural ecology of animals. Many birds, for example, display elaborate ornaments during courtship. These displays serve a dual purpose of signalling healthy or well-adapted individuals and good genes. The elaborate displays are driven by sexual selection as the displays serve as an advertisement of quality traits in sexual partners.[102]

Biogeography

As the name implies, biogeography is an amalgamation of the words *biology* and *geography*. The word was first coined by the German geographer, Friedrich Ratzel in 1891.[103] The Journal of Biogeography was established in 1974 and publishes "...papers dealing with all aspects of spatial, ecological and historical biogeography."[104]Biogeography and ecology share much of the same disciplinary roots. For example, the theory of island biogeography, elucidated by Robert MacArthur and E. O. Wilson in 1967[37] is considered one of the fundamentals of ecological theory.[105]

Biogeography has a long and rich history in the natural sciences where questions arise concerning the spatial distribution of plants and animals. Ecology and evolution provides the explanatory context for biogeographical studies.[103] Biogeographical patterns result from ecological processes that influences dispersal (or dispersion)[105] and from historical processes that split populations or species into different areas.[106] The biogeographic processes that result in the natural splitting of species explains much of the modern distribution of the Earth's biota. This area of focus is called vicariance biogeography and it is a sub-discipline of biogeography. It is a separate discipline because it specifically studies the branching, phylogenetic, or speciation process in evolutionary studies and explains much of the patterns in biodiversity across the globe.[106] [107] [108]

There are many applications in the field of biogeography that concern ecological systems and processes. For example, the range and distribution of biodiversity and invasive species responding to climate change is a serious concern and active area of research in context of global warming. [109] [110]

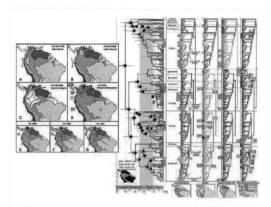

Biogeographers use genetics and palaeogeography to piece together the living history of organisms in their environments. This image shows the phylogeographic history of poison frogs living in South America. Maps to the left show the major biogeographic regions and paleogeographic changes that occured through time. The images to the right summarize the evolutionary history or relations among different frog species using a chronogram and phylogenetic trees derived from the molecular data collected from species that were sampled throughout the region.[111]

Molecular Ecology

There has long been an understanding of the important relationship between ecology and genetic inheritance.[2] This branch of research became more feasible with the development of genetic technologies, such as the polymerase chain reaction (PCR), and through the publication *Molecular Ecology* starting in 1992.[112] Molecular ecology uses various analytical techniques to study genes in evolutionary and ecological context. In 1994, professor John Avise played a leading role in popularizing this field of study through the publication of his book, *Molecular Markers, Natural History and Evolution* .[113] Newer genetic technologies made genetic sampling of organisms simpler and engendered a new and collaborative research paradigm that investigates and probes ecological questions that were otherwise intractable. Molecular ecology revealed previously obscured details in the intricacies of nature and improved resolution into probing questions about behavioural and biogeographical ecology. For example, molecular ecology revealed promiscuous sexual behaviour that is driven by female choice in pocket gophers [114] and multiple male partners in tree swallows previously thought to be socially monogamous.[115] In a biogeographical context, the marriage between genetics, ecology and evolution created a new sub-discipline called phylogeography.[116]

Ecology and the environment

The environment is external yet interlinked directly with ecology. Chemistry, temperature, pressure, gravity, energy, and sunlight are properties of Earth's environment that are relevant to ecology. Environmental and ecological relations are often studied through conceptually and practically manageable parts. However, once the effective environmental components are understood they conceptually link as a *holocoenotic*[117] system.

Ecology is often misused as a synonym for environment, but it differs from environmental studies, for example, because it is one of the few academic disciplines dedicated to holism.[12] The environment describes all factors and scales of study that are external to an organism, including abiotic factors such as temperature, radiation, light, chemistry, climate and geology, and biotic factors, including genes, cells, organisms, members of the same species (conspecifics) and other species that share a habitat.[118] In contrast, ecology focuses on biological relations and studies how these relate to the environment.[12] Ecosystem processes are consistent with the laws of thermodynamics. Armed with an understanding of metabolic and thermodynamic principles, a complete accounting of energy and material flow can be traced through an ecosystem.[119]

Metabolism and the early atmosphere

Metabolism − the rate at which energy and material resources are taken up from the environment, transformed within an organism, and allocated to maintenance, growth and reproduction − is a fundamental physiological trait.

—Ernst et al.[120] :991

The Earth's environment has remained in a dynamic equilibrium of ever-changing abiotic conditions, (temperature, humidity, etc.) and the atmosphere has changed significantly as a result of the gross metabolic activity of life. There is an evolving feedback loop between the ecological processes of life, geochemistry, and Earth's atmosphere.

Proceeding through early stages of life, major ecological transitions greatly modified Earth's geochemical cycles. The Earth formed approximately 4.5 billion years ago[121] and environmental conditions were too extreme for life to form for the first 500 million years. During this early Hadean period, the Earth started to cool allowing time for a crust and oceans to form. Environmental conditions were unsuitable for the origins of life until approximately 1 billion years after the Earth formed. The Earth's atmosphere transformed from hydrogen dominant, to one composed mostly of methane, and ammonia. Over the next billion years the metabolic activity of life transformed the atmosphere to higher concentrations of carbon dioxide, nitrogen, and water vapor. These gases changed the way that light from the sun hit the Earth's surface and greenhouse effects trapped in heat. There were untapped sources of free energy within the mixture of reducing and oxidizing gasses that set the stages for primitive ecosystems to evolve and, in turn, the atmosphere also evolved.[122]

The leaf is the primary site of photosynthesis in plants.

One of the earliest organisms was likely an anaerobic methanogen microbe that would have converted atmospheric hydrogen into methane ($4H_2 + CO_2 \rightarrow CH_4 + 2H_2O$). Anoxygenic photosynthesis converting hydrogen sulfide into other sulfur compounds or water ($2H_2S + CO_2 \rightarrow h\nu \rightarrow CH_2O \rightarrow H_2O \rightarrow + 2S$ or $2H_2 + CO_2 + h\nu \rightarrow CH_2O + H_2O$), as occurs in deep sea hydrothermal vents today, would have also reduced hydrogen and increased atmospheric methane. Early forms of fermentation would have also been a component of the primitive ecology producing higher levels of atmospheric methane. The transition to an oxygen dominant atmospheric transition did not begin until approximately 2.4-2.3 billion years ago, but photosynthetic processes had started 0.3 to 1 billion years prior. Hence, the transition to an oxygen environment was ecologically latent.[123] The evolution of the Earth's ecosystems demonstrates how smaller scale metabolic processes of life can regulate larger scale environmental phenomena, such as the Earth's atmosphere. This relationship has led to the development of the Gaia hypothesis, which states that there is a feedback process generated by living organisms that maintains the temperature of the Earth and atmospheric conditions within a narrow self-regulating range of tolerance. Hence, the gross ecology of the planet acts as a single regulatory or holistic unit called Gaia.[13]

Radiation: light, heat, and temperature

Almost all aspects of functional ecology are affected, directly or indirectly, by radiant energy from the sun. Different wavelengths of electromagnetic energy emanated by the sun provide inputs into the ecological energy budget of the planet. Radiant energy from the sun generates heat, provides photons of light measured as active energy in the chemical reactions of life, and also acts as a catalyst for genetic mutation.[2] [59] [119]

The biology of life operates within a certain range of temperatures. Heat is a form of energy that regulates temperature. Heat affects growth rates, activity, behavior and primary production. Temperature is largely dependent on the incidence of solar radiation. The latitudinal and longitudinal spatial variation of temperature greatly affects climates and consequently the distribution of biodiversity and levels of primary production in different ecosystems or biomes across the planet. Heat and temperature also relate importantly and differently affects two metabolic divisions in animals, poikilotherms, having a body temperature that is largely regulated and dependent on the temperature of the external environment, and homeotherms, having a body temperature that is internally regulated and maintained by expending metabolic energy.[2] [59] [119]

Light is the primary source of energy on the planet. Plants, algae, and some bacteria absorb light and assimilate the energy through photosynthesis. Organisms capable of assimilating energy by photosynthesis or through inorganic fixation of H_2S are autotrophs. Autotrophs are responsible for primary production and the assimilation of light energy that becomes metabolically stored as potentional energy in biochemical enthalpic bonds. Heterotrophs feed on autotrophs for their supply of energy and nutrients. Hence, there is a relationship between light, production, and supplies of energy that affects the distribution, composition and structure of ecosystem dynamics across the planet.[2] [59] [119]

Physical environments

Water

The rate of diffusion of carbon dioxide and oxygen is approximately 10,000 times slower in water than it is in air. When soils become flooded, they quickly loose oxygen from low-concentration (hypoxic) to an (anoxic) environment where anaerobic bacteria thrive among the roots[124]. Aquatic plants exhibit a wide variety of morphological and physiological adaptations that allow them to survive, compete and diversify these environments. For example, the roots and stems develop large cellular air spaces to allow for the efficient transportation gases (for example, CO_2 and O_2) used in respiration and photosynthesis. In drained soil, microorganisms use oxygen during respiration. In aquatic environments, anaerobic soil microorganisms use nitrate, manganic ions, ferric ions, sulfate, carbon dioxide and some organic compounds. The activity of soil microorganisms and the chemistry of the water reduces the oxidation-reduction potentials of the water. Carbon dioxide, for example, is reduced to methane (CH_4) by methanogen bacteria. Salt water also requires special physiological adaptations to deal with water loss. Salt water plants (or halophytes) are able to osmo-regulate their internal salt (NaCl) concentrations or develop special organs for shedding salt away.[124]. The physiology of fish is also specially adapted to deal with high levels of salt through osmoregulation. Their gills form electrochemical gradients that mediate salt excrusion in salt water and uptake in fresh water.[125]

Wetland conditions such as shallow water, high plant productivity, and anaerobic substrates provide a suitable environment for important physical, biological, and chemical processes. Because of these processes, wetlands play a vital role in global nutrient and element cylcles.:29[124]

Gravity

The shape and energy of the land is affected to a large degree by gravitational forces. On a larger scale, the distribution of gravitational forces on the earth are uneven and influence the shape and movement of tectonic plates as well as having an influence on geomorphic processes such as orogeny and erosion. These forces govern some of the geo-physical properties and distributions of biomes across the Earth. On a organism scale, gravitational forces provide directional cues for plant and fungal growth (gravitropism), orientation cues for animal migrations, and influences the biomechanics and size of animals.[2]

Pressure

Pressure effects the environment and the organism. It acts as a mechanical force with close connections to gravity causing increased levels of pressure moving toward the Earth. Pressure exerts significant influence over the atmosphere, climate, water environments, and on smaller scale there are osmotic forces at work. Organisms are physiologically sensitive and adapted to atmospheric and osmotic water pressures.[2] Water transportation through trees, for example, is an important eco-physiological parameter.[126] [127] Water pressure in the depths of oceans requires adaptations to deal with the different living conditions. Mammals, such as whales, dolphins and seals require special adaptations to deal with the change in sound due to water pressure differences.[128] Climatic and osmotic pressure places physiological constraints on organisms, such as flight and respiration at high altitudes, or diving to deep ocean depths. These constraints influence vertical limits of ecosystems in the biosphere.[2]

Wind and turbulence

Turbulent forces in air and water have significant effects on the environment and ecosystem distribution, form and dynamics. On a planetary scale, ecosystems are affected by circulation patterns in the global trade winds. Locally, wind power and the turbulent forces it creates can influence heat, nutrient, and biochemical profiles of ecosystems.[2]

For example, wind running over the surface of a lake creates turbulence, mixing the water column and influencing the environmental profile to create thermally layered zones, partially governing how fish, algae, and other parts of the aquatic ecology are structured.[129] [130]

Wind speed and turbulence also exert influence on rates of evapotranspiration rates and energy budgets in plants and animals [124] [131]

Fire

Plants take in carbon dioxide, stripping and emitting oxygen into the atmosphere. Approximately 350 million years ago (near the Devonian period) the photosynthetic process brought atmospheric oxygen levels above 17% in concentration, which allowed for the combustion of fire.[132] Fire releases CO_2 and converts fuel into ash and tar. Fire is a significant ecological parameter that raises many issues pertaining to its control and suppression in management.[133] While the issue of fire in relation to ecology and plants has been recognized for a long time[134], Charles Cooper brought attention to the issue of forest fires in relation to the ecology of forest fire suppression and management in the 1960s.[135] [136] The association for fire ecology launched a journal, "Fire Ecology", in 2005 that is specifically devoted to the study of fire ecology and management.[137].

Forest fires modify the land leaving behind an environmental mosaic of differing ecological niche's (left). Some species are adapted to forest fires, such as pine trees that open their cones only after fire exposure (right).

Fire creates environmental mosaics and a patchiness to ecosystem age and canopy structure. Native North Americans were among the first to influence fire regimes by controlling their spread near their homes or by lighting fires to stimulate the production of herbaceous foods and basketry materials.[138] Most ecosystem are adapted to some level of natural fire cycles. Plants, for example, are equipped with a variety of special adaptations to deal with forest fires. The altered state of soil nutrient supply and cleared canopy structure creates a new niche for seedling establishment.[139] [140] Some species (e.g., *Pinus halepensis*) cannot germinate until after their seeds have lived through a fire. This environmental trigger for seedlings is called serotiny.[141] Some compounds from smoke also promote seed germination.[142]

Biogeochemistry

Ecologists study and measure nutrient budgets to understand how these materials are regulated and flow through the environment.[2] [59] [119] This research has led to an understanding that there is a global feedback between ecosystems and the physical parameters of this planet including minerals, soil, pH, ions, water and atmospheric gases. There are six major elements, including H (hydrogen), C (carbon), N (nitrogen), O (oxygen), S (sulfur), and P (phosphorus) that form the constitution of all biological macromolecules and feed into the Earth's geochemical processes. From the smallest scale of biology the combined effect of billions upon billions of ecological processes amplify and ultimately regulate the biogeochemical cycles of the Earth. Understanding the relations and cycles mediated between these elements and their ecological pathways has significant bearing toward understanding global biogeochemistry.[143]

The ecology of global carbon budgets gives one example of the linkage between biodiversity and biogeochemistry. For starters, the ocean is estimated to hold 40,000 Gt carbon, vegetation and soil is estimated to hold 2070 Gt carbon, and fossil fuel emissions are estimated to emit an annual flux of 6.3 Gt carbon.[144] At different times in the Earth's history there has been major restructuring in these global carbon budgets that was regulated to a large extent by the ecology of the land. For example, through the early-mid Eocene volcanic out gassing, the oxidation of methane stored in wetlands, and seafloor gases increased atmospheric CO_2 concentrations to levels as high as 3500 ppm.[145] In the Oligocene, from 25 to 32 million years ago, there was another significant restructuring in the global carbon cycle as grasses evolved a special type of C4 photosynthesis and expanded their ranges. This new photosynthetic pathway evolved in response to the drop in atmospheric CO_2 concentrations below 550 ppm.[146] Ecosystem functions such as these feed back significantly into global atmospheric models for carbon cycling. Loss in the abundance and distribution of biodiversity causes global carbon cycle feedbacks that are expected to increase rates global warming in the next century.[147] Global warming melting large sections of permafrost creates a new mosaic of flooded areas where decomposition emits methane (CH_4). Hence, there is a relationship between global warming, decomposition and respiration in soils and wetlands producing significant climate feedbacks and alters global biogeochemical cycles.[148] [149] There is concern over methane increases in the atmosphere in context of the carbon cycle, because methane is also a greenhouse gas that is 23 times more effective at absorbing long-wave radiation on a 100 year time scale.[150]

Historical roots of ecology

In the early 20th century, ecology was called scientific natural history and was influenced by the analytical precision of Newtonian sciences.[151] A comprehensive historical account of ecology is a complicated task because ecology is one of the most diverse of the scientific disciplines.[152] Several published books provide extensive coverage of the classics.[153] [154] The term "ecology" (German: *Oekologie*) is a more recent scientific development and was first coined by the German biologist Ernst Haeckel in his book *Generelle Morpologie der Organismen* (1866). The definition offered by Haeckel appeared in the frontispiece of the classical text *Principles of Animal Ecology.*[2]

By ecology we mean the body of knowledge concerning the economy of nature-the investigation of the total relations of the animal both to its inorganic and its organic environment; including, above all, its friendly and inimical relations with those animals and plants with which it comes directly or indirectly into contact-in a word, ecology is the study of all those complex interrelations referred to by Darwin as the conditions of the struggle of existence.

—Haeckel's definition quoted in Esbjorn-Hargens[155] :6

Ernst Haeckel (left) and Eugenius Warming (right), two early founders of ecology.

Opinions differ on who was the founder of modern ecological theory. Some some mark Haeckel's definition as the beginning[156], others suggest it started with the Greeks, such as Aristotle and his student Theoprastis[157], some say it was Eugen Warming[158] and others posit that the science of ecology began with Carl Linnaeus' research principals on the economy of nature that matured in the early 18th century.[159] [49] The works of Carl Linnaeus influenced Darwin, for example, as evidenced by his reference to ecology through his adopted usage of Linnaeus' phrase *economy or polity of nature* in The Origin of Species.[160] Ernst Haeckel was strongly influenced by Darwin's work. He defined ecology in reference to the economy of nature and this has lead some to question if ecology is synonymous with Linnaeus' concepts for the economy of nature.[159] Some have suggested that ecology started with Alexander von Humbolt (1809-1882), who was also admired by Charles Darwin.[151] Baron Humbolt was among the first to recognize ecological gradients and alluded to the modern law of species to area relationships in ecology.[161] [162]

The modern synthesis of ecology is a young science that flourished and attracted much research attention around the same time as evolutionary studies at the end of the 19th century. However, many observations, interpretations and discoveries relating to ecology extend back to much earlier studies in natural history. For example, the concept on the balance or regulation of nature can be traced back to Herodotos (died c. 425 BC) who described an early account of mutualism along the Nile river where crocodiles open their mouths to beneficially allow sandpipers safe access to pluck leaches away.[152] . In the broader contributions to the historical development of the ecological sciences, Aristotle is considered one of the earliest naturalists who had a highly influential role in the philosophical development of ecological sciences. One of Aristotle's students, Theophrastus, made astute ecological observations about plants and posited a philosophical stance about the autonomous relations between plants and their environment that is more in line with modern ecological thought. Both Aristotle and Theophrastus made extensive observations on plant and animal migrations, biogeography, physiology, and their habits in what might be considered a modern analog of the ecological niche.[163] [157]

Carl Linnaeus (1707–1778), a well known naturalist also holds a prominent place in the history of ecological sciences as he invented the first branch of ecological study he called the economy of nature.[49] Linnaeus was one of the first to attempt to define on the balance of nature, which had previously been held as an assumption rather than formulated as a testable hypothesis. From Aristotle to Darwin, however, the natural world was predominantly considered static and unchanged since its original creation. Hence, there was little appreciation and understanding for the dynamic and reciprocal relations between organisms, their adaptations and modifications to the environment.[164] [155]

Nowhere can one see more clearly illustrated what may be called the sensibility of such an organic complex,--expressed by the fact that whatever affects any species belonging to it, must speedily have its influence of some sort upon the whole assemblage. He will thus be made to see the impossibility of studying any form completely, out of relation to the other forms,--the necessity for taking a comprehensive survey of the whole as a condition to a satisfactory understanding of any part.

—Stephen Forbes (1887)[165]

While Charles Darwin is most notable for his treatise on evolution[166], he was also a notable and astute ecologist as he meticuously researched earthworms in relation to soil ecology[167] and in The Origin of Species he made note of the very first ecological experiment that was published in 1816.[168] [169] In the science leading up to Darwin the notion of evolving species was gaining popular support. This scientific paradigm changed the way that researchers approached the ecological sciences.

The first American ecology book was published in 1905 by Frederic Clements.[170] Frederic Clements forwarded the idea of plant communities as a superorganism. According to this premise, single species populations could be classified, using identifiable plant associations, into larger super-organism entities that were believed to progress through regular and determined stages of seral development. Seral stages of an ecosystem were thought to be analogous to developmental stages of an organism. Not until the 1970's had the Clementsian paradigm been overthrown by the Gleasonian paradigm [171] which emphasized the overriding role of individual organisms and their life histories in the development of community associations. Hence, seral stages of ecosystems were not a predictable and reproduceable outcome. Each ecological community develops from the unique and coincidental juxtopostion of organism. The Gleasonian paradigm suggests that a precise structural uniformity to ecosystems does not exist because the different community associations are historically and geographically context dependent.[172] According to Gleason[172] , there is much variation among sites that is driven by extrinsic, not internal factors.

The number of authors publishing on the topic of ecology has grown considerably since the turn of 20th century.[173] The explosion of information that is now available to the modern researcher of ecology makes it is an impossible task for any individual to sift through the entire history. The identification of classics in the history of ecology is a difficult designation to make.[174]

Ellen Swallow Richards, the first female student and instructor at MIT.

"Vito Volterra's "Variations and Fluctuations of the Number of Individuals in Animal Species Living Together" would be so recognized by population ecologists. R. L. Lindeman's, "The Trophic-Dynamic Aspect of Ecology"; A. S. Watt's "Pattern and Process in the Plant Community"; G. E. Hutchinson's ambiguously titled "Concluding Remarks" (concealing his multidimensional formulation of the niche concept), and Robert MacArthur's "On the Relative Abundance of Bird Species" (including his broken-stick model) have all been described as classic, and many persons would agree. The designation of an article as a classic of ecology, however, may or may not represent a consensus and may be attributable to the idiosyncrasy of the designator."[174] :32

Political persuasion, subjectivity and other overriding social factors are also encountered in the citation history of ecological journals.[175] Feminist researchers, such as Ellen Swallow Richards, might not be as prominently noted as their male counterparts.[176] Researchers are studying what factors contribute to the citation rates of certain papers and have published selected lists of papers that identify significant contributions to ecology.[175] [174]

"...social factors, such as the professional standing of the cited author, play a significant role in citation decisions in ecology. Furthermore, the dependence of the citation rates of ecological papers on the direction of study outcome with respect to the hypothesis tested suggests that citations in ecological papers are used as rhetorical devices to convince the readers of the validity of the study claims rather than as simple acknowledgements of the sources of background information."[175] :31

Ecology has also developed in other nations, including Russia's Vladimir Vernadsky and his development of the biosphere concept in the 1920's[177] or Japan's Kinji Imanishi and his concepts of harmony in nature and habitat segregration in the 1950's[178] where recognition or scientific importance of the contributions to ecology is hampered by language and translation barriers.[177] The history of ecology remains an active area of study.

A list of founders, innovators and their significant contributions to ecology, from Romanticism onward.

Notable figure	Lifespan	Major contribution & citation
Antoni van Leeuwenhoek	1632-1723	First to develop concept of food chains [158]
Carl Linnaeus	1707–1778	Influential naturalist, inventor of science on the economy of nature [49] [159]
Alexander Humboldt	1769–1859	First to describe ecological gradient of latitudinal biodiversity increase toward the tropics[179] in 1807 [162]
Charles Darwin	1809-1882	Founder of evolution by means of natural selection, founder of ecological studies of soils [166] [167]
Herbert Spencer	1820–1903	Early founder of social ecology, coined the phrase 'survival of the fittest' [159] [180]
Karl Möbius	1825-1908	First to develop concept of ecological community, biocenosis, or living community [181] [182] [183]
Ernst Haeckel	1834-1919	Invented the term ecology, popularized research links between ecology and evolution [155] [160]
Victor Hensen	1835-1924	Invented term plankton, developed quantitative and statistical measures of productivity in the seas [160]
Eugenius Warming	1841-1924	Early founder of Ecological Plant Geography [184]
Ellen Swallow Richards	1842–1911	Pioneer and educator who linked urban ecology to human health [176] [185]
Stephen Forbes	1844–1930	Early founder of entomology and ecological concepts in 1887[186] [187] [165]
Vladimir Vernadsky	1869-1939	Founded the biosphere concept [177]
Henry C. Cowles	1869-1939	Pioneering studies and conceptual development in studies of ecological succession [188]
Arthur G. Tansley	1871–1955	First to coin the term ecosystem in 1936 and notable researcher [189] [182] [190]
Charles Christopher Adams	1873-1955	Animal ecologist, biogeographer, author of first American book on animal ecology in 1913, founded ecological energetics [191] [192]
Frederic Clements	1874-1945	Authored the first influential American ecology book in 1905 [171]
Victor Ernest Shelford	1877-1968	Founded physiological ecology, pioneered food-web and biome concepts, founded The Nature Conservancy [193] [194]
Alfred J. Lotka	1880-1949	First to propose mathematical model explaining trophic (predator-prey) interactions using logistic equation [34]
Henry Gleason	1882-1975	Early ecology pioneer, quantitative theorist, author, and founder of the individualistic concept of ecology [195] [171]
Charles S. Elton	1900-1991	'Father' of animal ecology, pioneered food-web & niche concepts and authored influential *Animal Ecology* text [196] [193]
G. Evelyn Hutchinson	1903-1991	Limnologist and conceptually advanced the niche concept [197] [198] [199]
Eugene P. Odum	1913-2002	Co-founder of ecosystem ecology and ecological thermodynamic concepts [190] [193] [200] [201]
Howard T. Odum	1924–2002	Co-founder of ecosystem ecology and ecological thermodynamic concepts [190] [193] [202] [203] [200] [201]
Robert MacArthur	1930–1972	Co-founder on Theory of Island Biogeographer and innovator of ecological statistical methods [204]

Ecosystem services and the biodiversity crisis

Ecosystems regulate the global geophysical cycles of energy, climate, soil nutrients, and water that in turn support and grow natural capital (including the environmental, physiological, cognitive, cultural, and spiritual) dimensions of life. Ecosystems are considered common-pool resources because ecosystems do not exclude beneficiaries and they can be depleted or degraded.[205] For example, green space within communities provides common-pool health services. Research shows that people who are more engaged with regular access to natural areas have lower rates of diabetes, heart disease and psychological disorders.[206] These ecological health services are regularly depleted through urban development projects that do not factor in the common-pool value of ecosystems.[207] [208] The ecological commons delivers a diverse supply of community services that sustains the well-being of human society.[209] [210] The Millineum Ecosystem

Pollination by a bumblebee, a type of ecosystem service

Assessment[211], an international UN initiative involving more than 1,360 experts worldwide, identifies four main ecosystem service types having 30 sub-categories stemming from natural capital. The ecological commons includes provisioning (e.g., food, raw materials, medicine, water supplies), regulating (e.g., climate, water, soil retention, flood retention), cultural (e.g., science and education, artistic, spiritual), and supporting (e.g., soil formation, nutrient cycling, water cycling) services.

Policy and human institutions should rarely assume that human enterprise is benign. A safer assumption holds that human enterprise almost always exacts an ecological toll - a debit taken from the ecological commons.[212] :95

Ecology is an economic science that uses many of the same terms and methods that are used in accounting.[213] There is a journal called *Ecological Economics*[[214]] and the *International Society of Ecological Economics* [[215]] that researches and publishes on this part of the global economy. Natural capital is the stock of materials or information stored in biodiveristy that generates services that can enhance the welfare of communities.[216] Population losses are the more sensitive indicator of natural capital than are species extinction in the accounting of ecosystem services. The prospect for recovery in the economic crisis of nature is grim. Populations, such as local ponds and patches of forest are being cleared away and lost at rates that exceed species extinctions.[217]

This paper uses the concepts of human carrying capacity and natural capital to argue that prevailing economic assumptions regarding urbanization and the sustainability of cities must be revised in light of global ecological change. While we are used to thinking of cities as geographically discrete places, most of the land "occupied" by their residents lies far beyond their boders. The total area of land required to sustain an urban region (its "ecological footprint") is typically at least an order of magnitude greater than that contained within municipal boundaries or the associated built-up area.[218] :121

The WWF 2008 living planet report [219] and other researchers[220] report that human civilization has exceeded the bio-regenerative capacity of the planet. This means that human consumption is extracting more natural resources than can be replenished by ecosystems around the world. In 1992, professor William Rees developed the concept of our ecological footprint. The ecological footprint is a way of accounting the level of impact that human development is having on the Earth's ecosystems.[218] All indications are that the human enterprise is unsustainable as the ecological footprint of society is placing too much stress on the ecology of the planet.[220] The mainstream growth based economic system adopted by governments worldwide does not include a price or markets for natural capital. This type of economic system places further ecological debt onto future generations.[221]

Human societies are increasingly being placed under stress as the ecological commons is diminished through an accounting system that has incorrectly assumed "...that nature is a fixed, indestructible capital asset." [222] :44 While nature is resilient and it does regenerate, there are limits to what can be extracted, but conventional monetary analyses are unable to detect the problem.[223] [224] Evidence of the limits in natural capital are found in the global assessments of biodiversity, which indicate that the current epoch, the Anthropocene[225] , has entered the sixth great extinction period.[226] The ecology of the planet has been radically transformed by human society and development causing massive loss of ecosystem services that otherwise deliver and freely sustain equitable benefits to human society through the ecological commons. The ecology of the planet is further threatened by global warming, but investments in nature conservation can provide a regulatory feedback to store and regulate carbon and other greenhouse gases.[227] [228] The field of conservation biology involves ecologists that are researching the nature of the biodiversity threat and searching for solutions to sustain the planets ecosystems for future generations.[229]

> "Human activities are associated directly or indirectly with nearly every aspect of the current extinction spasm."[226] :11472

The current wave of threats, including massive extinction rates and concurrent loss of natural capital to the detriment of human society is happening rapidly. This is called a biodiversity crisis because 50% of the worlds species are predicted to go extinct within the next 50 years.[230] [231] The worlds fisheries are facing dire challenges as the threat of global collapse appears immenent and with serious ramifications for the well-being of humanity[232] Governments of the G8 met in 2007 and set forth 'The Economics of Ecosystems and Biodiversity' (TEEB) initiative[233]:

> In a global study we will initiate the process of analyzing the global economic benefit of biological diversity, the costs of the loss of biodiversity and the failure to take protective measures versus the costs of effective conservation.[234]

Ecologists are teaming up with economists to measure the wealth of ecosystems and to express their value as a way of finding solutions to the biodiversity crisis.[235] [236] [237] Some researchers have attempted to place a dollar figure on ecosystem services, such as the value of the Canadian Boreal Forest contributing services such as carbon storage is estimated at US$3.7 trillion. The annual value for ecological services oof the Boreal Forest is estimated at US$93.2 billion, or 2.5 greater than the annual value of resource extraction. These ecological economic values are not currently included in calculations of national income accounts or the GDP.[238] [239]

See also

- Acoustic ecology
- Agroecology
- Biodiversity
- Biotope
- Biogeography
- Climate
- → Conservation movement
- Earth science
- Ecoacoustics
- Ecohydrology
- → Ecological economics
- Ecological Forecasting
- Ecology movement
- Ecology of contexts
- Ecosystem
- Ecosystem model
- Ecological Relationships
- Ecosystem services
- Ecotope
- ELDIS, a database on ecological aspects of economical development.

Bachalpsee in the Swiss Alps; generally mountainous areas are less affected by human activity.

- Environment
- Forest farming
- Forest gardening
- Habitat conservation
- Human ecology
- Industrial ecology
- Knowledge ecology
- Landscape ecology
- Landscape limnology
- Molecular ecology
- Natural capital
- Natural resource
- Natural resource management
- Nature
- Palaeoecology
- Social ecology
- Sustainability
- Sustainable development

Lists

- Index of biology articles
- Glossary of ecology
- List of ecologists
- List of important publications in biology#Ecology
- Outline of biology

References

- Campbell, Neil A.; Brad Williamson; Robin J. Heyden (2006). *Biology: Exploring Life* [240]. Boston, Massachusetts: Pearson Prentice Hall. ISBN 0132508826.
- Brinson, M. M., Lugo, A. E. and Brown, S. (1984). Primary Productivity, Decomposition and Consumer Activity in Freshwater Wetlands. Annual Review of Ecology and Systematics, 12, 123-161.
- David, R. D. and Welsh, H. H. (2004). On the ecological role of salamanders. Annual Review of Ecology and Systematics, 35, 405-434
- Gould, S. J. and Lloyd, E. A. (1999). Individuality and adaptation across levels of selection: How shall we name and generalize the unit of Darwinism? Proceedings of the National Academy of Science, 96(21), 11904-11909.[241]
- Haeckel, E. (1866) *General Morphology of Organisms; General Outlines of the Science of Organic Forms based on Mechanical Principles through the Theory of Descent as reformed by Charles Darwin.* Berlin
- Lovelock, J. (2003). Gaia: The living earth. Nature, 426, 769-770 [242]
- Odum, E. P. (1971) *General Principles of Ecology, Third Edition* W. B. Suanders Company. pp 17–20
- Odum, E. P. (1977) The emergence of ecology as a new integrative discipline. Science, 195, 1289-1293.
- Warming, E. (1909) Oecology of Plants - an introduction to the study of plant-communities. Clarendon Press, Oxford.
- Whiles, M. R., Lips, K. R., Pringle, C. M., Kilham, S. S., Bixby, R. J., Brenes, R., Connelly, S., Colon-Gaud, J. C., Hunte-Browjn, M., Huryn, A. D., Montgomery, C., and Peterson, S. 2006. The effects of amphibian population declines on the structure and function of Neotropical stream ecosystems. Front Ecol Environ, 4(1),

27–34

External links

- stanford.edu/entries/ecology/ Ecology (Stanford Encyclopedia of Philosophy) [243]
- Science Aid: Ecology [244] High School (GCSE, Alevel) Ecology.
- Ecology Journals [245] List of scientific journals related to Ecology
- Ecology Dictionary - Explanation of Ecological Terms [246]
- [78] International Society for Behavioral Ecology

References

[1] Begon, M.; Townsend, C. R., Harper, J. L. (2006). *Ecology: From individuals to ecosystems. (4th ed.)*. Blackwell. ISBN 1405111178.

[2] Allee, W.; Emerson, A. E., Park, O., Park, T., and Schmidt, K. P. (1949). *Principles of Animal Ecology*. W. B. Saunders Company. ISBN 0721611206.

[3] Smith, R.; Smith, R. M. (2000). *Ecology and Field Biology. (6th ed.)*. Prentice Hall. ISBN 0321042905.

[4] Omerod, S.J.; Pienkowski, M.W.; Watkinson, A.R. (1999). "Communicating the value of ecology". *Journal of Applied Ecology* **36**: 847–855.

[5] Phillipson, J.; Lowe, P.; Bullock, J.M. (2009). "Navigating the social sciences: interdisciplinarity and ecology". *Journal of Applied Ecology* **46**: 261–264.

[6] Steward T. A. Pickett, Mary L. Cadenasso, J. Morgan Grove, Peter M. Groffman, Lawrence E. Band, Christopher G. Boone, William R. Burch Jr., C. Susan B. Grimmond, John Hom, Jennifer C. Jenkins, Neely L. Law, Charles H. Nilon, Richard V. Pouyat, Katalin Szlavecz, Paige S. Warren, Matthew A. Wilson (2008). "Beyond Urban Legends: An Emerging Framework of Urban Ecology, as Illustrated by the Baltimore Ecosystem Study". *BioScience* **58**: 139–150.

[7] Aguirre, A.A. (2009). "Biodiversity and Human Health". *EcoHealth*. doi: 10.1007/s10393-009-0242-0 (http://dx.doi.org/10.1007/s10393-009-0242-0).

[8] http://www.lternet.edu/

[9] http://www.hubbardbrook.org/

[10] Silverton, J.; Poulton, P.; Johnston, E.; Grant, E.; Heard, M.; Biss, P. M. (2006), " The Park Grass Experiment 1856–2006: its contribution to ecology (http://www.demonsineden.com/Site/Research_publications_files/Silvertown et al.2006.pdf)", *Journal of Ecology* **94** (4): 801-814,

[11] Schneider, D. D. (2001), "[The Rise of the Concept of Scale in Ecology The Rise of the Concept of Scale in Ecology]", *BioScience* **51** (7): 545–553, The Rise of the Concept of Scale in Ecology

[12] Odum, E. P. (1977). "The emergence of ecology as a new integrative discipline". *Science* **195**: 1289–1293.

[13] Lovelock, J. (2003). "The living Earth". *Nature* **426** (6968): 769–770. doi: 10.1038/426769a (http://dx.doi.org/10.1038/426769a). PMID 14685210 (http://www.ncbi.nlm.nih.gov/pubmed/14685210).

[14] Nachtomy, Ohad; Shavit, Ayelet; Smith, Justin (2002), " Leibnizian organisms, nested individuals, and units of selection (http://www.springerlink.com/content/25625863427113r0/)", *Theory in Biosciences* **121** (2),

[15] Begon, M.; Townsend, C. R.; Harper, J. L. (2006), *Ecology: From Individuals to Ecosystems* (http://books.google.ca/books?id=Lsf1lkYKoHEC&printsec=frontcover&dq=ecology&lr=&as_drrb_is=b&as_minm_is=0&as_miny_is=2004&as_maxm_is=0&as_maxy_is=2009&as_brr=0&client=firefox-a&cd=1#v=onepage&q=&f=false) (4th ed.), Oxford, UK: Blackwell Publishing, ISBN 978-1-4051-1117-1,

[16] Zak, K. M.; Munson, B. H. (2008), " An Exploratory Study of Elementary Preservice Teachers' Understanding of Ecology Using Concept Maps. (http://www.duluth.umn.edu/~kgilbert/ened5560-1/Readings/SciEd-JEESpring2008-ZakMunsonArticleUpdated.pdf)", *The Journal of Environmental Education* **39** (3): 32–46,

[17] Edward O.Wilson, editor, Frances M.Peter, associate editor, *Biodiversity*, National Academy Press, March 1988 ISBN 0-309-03783-2 ; ISBN 0-309-03739-5 (pbk.), online edition (http://darwin.nap.edu/books/0309037395/html/R2.html)

[18] Noss, R.; Cooperrider, A. (1994), *Saving Natures Legacy: Protecting and Restoring Biodiversity*, Washington, DC: Island Press

[19] Margulis, Lynn (1992). " Biodiversity: molecular biological domains, symbiosis and kingdom origins (http://www.sciencedirect.com/science?_ob=ArticleURL&_udi=B6T2K-49NY23W-65&_user=10&_rdoc=1&_fmt=&_orig=search&_sort=d&_docanchor=&view=c&_searchStrId=1142524797&_rerunOrigin=scholar.google&_acct=C000050221&_version=1&_urlVersion=0&_userid=10&md5=5993d10eb706481d7d7358a3dd9e06c8)". *Biosystems* **28** (1-3): 107–108. .

[20] Wiens, J. J.; Graham, C. H. (2005), " Integrating Evolution, Ecology, and Conservation Biology (http://life.bio.sunysb.edu/ee/grahamlab/pdf/Wiens_Graham_AnnRev2005.pdf)", *Annual Review of Ecology, Evolution, and Systematics* **36**: 519–539,

[21] Hutchinson, G. E. (1957). *A Treatise on Limnology.*. New York: Wiley & Sons.. pp. 1015. ISBN 0471425729.

[22] Hutchinson, G. E. (1957). " Concluding remarks. (http://symposium.cshlp.org/content/22/415.full.pdf+html)". *Cold Spring Harb Symp Quant Biol* **22**: 415–427. .

[23] McGill, B. J.; Enquist, B. J.; Weiher, E.; Westoby, M. (2006). "Rebuilding community ecology from functional traits". *Trends in Ecology and Evolution* **21** (4): 178–185.

[24] Hardin, G. (1960). "The competitive exclusion principal.". *Science* **131** (3409): 1292–1297. doi: 10.1126/science.131.3409.1292 (http://dx.doi.org/10.1126/science.131.3409.1292).

[25] Whittaker, R. H.; Levin, S. A.; Root, R. B. (1973). " Niche, Habitat, and Ecotope (http://www.jstor.org/stable/2459534?seq=6)". *The American Naturalist* **107** (955): 321–338. .

[26] Hastings, A. Byers, J. E., Crooks, J. A., Cuddington, K., Jones, C. J., Lambrinos, J. G., Talley, T. S. and Wilson, W. G., A; Byers, JE; Crooks, JA; Cuddington, K; Jones, CG; Lambrinos, JG; Talley, TS; Wilson, WG (2007). "Ecosystem engineering in space and time". *Ecology Letters* **10** (2): 153–164. doi: 10.1111/j.1461-0248.2006.00997.x (http://dx.doi.org/10.1111/j.1461-0248.2006.00997.x). PMID 17257103 (http://www.ncbi.nlm.nih.gov/pubmed/17257103).

[27] Jones, Clive G.; Lawton, John H.; Shachak, Moshe (1994). "Organisms as ecosystem engineers". *Oikos* **69** (3): 373–386. doi: 10.2307/3545850 (http://dx.doi.org/10.2307/3545850).

[28] Wright, J.P.; Jones, C.G. (2006). "The Concept of Organisms as Ecosystem Engineers Ten Years On: Progress, Limitations, and Challenges". *BioScience* **56**: 203–209. doi: 10.1641/0006-3568(2006)056[0203:TCOOAE]2.0.CO;2 (http://dx.doi.org/10.1641/0006-3568(2006)056[0203:TCOOAE]2.0.CO;2).

[29] Laland, K. N.; Odling-Smee, F.J.; Feldman, M.W. (1999). " Evolutionary consequences of niche construction and their implications for ecology (http://www.pubmedcentral.nih.gov/articlerender.fcgi?tool=pmcentrez&artid=17873)". *PNAS* **96** (18): 10242–10247. doi: 10.1073/pnas.96.18.10242 (http://dx.doi.org/10.1073/pnas.96.18.10242). PMID 10468593 (http://www.ncbi.nlm.nih.gov/pubmed/10468593).

[30] http://www.springerlink.com/content/1438-3896?sortorder=asc&p=93932389f9764a2aadcbe167b466fcef&o=0

[31] Turchin, P. (2001), "Does Population Ecology Have General Laws?", *Oikos* **94** (1): 17–26

[32] Vandermeer, J. H.; Goldberg, D. E. (2003), *Population ecology: First principles*, Woodstock, Oxfordshire: Princeton University Press, ISBN 0-691-11440-4

[33] Johnson, J. B.; Omland, K. S. (2004), " Model selection in ecology and evolution. (http://www.usm.maine.edu/bio/courses/bio621/model_selection.pdf)", *Trends in Ecology and Evolution* **19** (2): 101–108,

[34] Berryman, A. A. (1992). "The Origins and Evolution of Predator-Prey Theory". *Ecology* **73** (5): 1530–1535.

[35] Terms and definitions directly quoted from: Wells, J. V.; Richmond, M. E. (1995). " Populations, metapopulations, and species populations: What are they and who should care? (http://www.uoguelph.ca/zoology/courses/BIOL3130/wells11.pdf)". *Wildlife Society Bulletin* **23** (3): 458-462. .

[36] Whitham, T. G. (1978). "Habitat Selection by Pemphigus Aphids in Response to Response Limitation and Competition". *Ecology* **59** (6): 1164–1176.

[37] MacArthur, R.; Wilson, E. O. (1967), *The Theory of Island Biogeography*, Princeton, NJ: Princeton University Press

[38] Pianka, E. R. (1972). "r and K Selection or b and d Selection?". *The American Naturalist* **106** (951): 581–588.

[39] Levins, R. (1969). " Some demographic and genetic consequences of environmental heterogeneity for biological control. (http://books.google.ca/books?hl=en&lr=&id=8jfmor8wVG4C&oi=fnd&pg=PA162&ots=GJCtM8hhbu&sig=kSiFKPIaX_p_ZCeQZtf1G0k4ib4#v=onepage&q=&f=false)". *Bulletin of the Entomological Society of America* **15**: 237-240. .

[40] Levins, R. (1970). Gerstenhaber, M.. ed. *Extinction. In: Some Mathematical Questions in Biology* (http://books.google.ca/books?id=CfZHU1aZqJsC&dq=Some+Mathematical+Questions+in+Biology&printsec=frontcover&source=bl&ots=UXQZc5WZwK&sig=1F6yBuo09HOAwxFL5QA8Ak_BLA0&hl=en&ei=V2U9S5SyLpDflAe40qmdBw&sa=X&oi=book_result&ct=result&resnum=1&ved=0CAwQ6AEwAA#v=onepage&q=&f=false). pp. 77-107. .

[41] Hanski, I. (1998). " Metapopulation dynamics (http://www.helsinki.fi/~ihanski/Articles/Nature 1998 Hanski.pdf)". *Nature* **396**: 41-49. .

[42] Hanski, I.; Gaggiotti, O. E., eds (2004). *Ecology, genetics and evolution of metapopulations.* (http://books.google.ca/books?id=EP8TAQAAIAAJ&q=ecology,+genetics,+and+evolution+of+metapopulations&dq=ecology,+genetics,+and+evolution+of+metapopulations&cd=1). Burlington, MA: Elsevier Academic Press. ISBN 0-12-323448-4. .

[43] Johnson, M. T.; Strinchcombe, J. R. (2007). "An emerging synthesis between community ecology and evolutionary biology.". *Trends in Ecology and Evolution* **22** (5): 250–257.

[44] Brinson, M. M.; Lugo, A. E.; Brown, S (1981). "Primary Productivity, Decomposition and Consumer Activity in Freshwater Wetlands". *Annual Review of Ecology and Systematics* **12**: 123–161. doi: 10.1146/annurev.es.12.110181.001011 (http://dx.doi.org/10.1146/annurev.es.12.110181.001011).

[45] Davic, R. D.; Welsh, H. H. (2004). "On the Ecological Role of Salamanders". *Annual Review of Ecology and Systematics* **35**: 405–434.

[46] Paine, R. T. (1980), " Food Webs: Linkage, Interaction Strength and Community Infrastructure (http://www.jstor.org/stable/4220)", *Journal of Animal Ecology* **49** (3): 667–685,

[47] Abrams, P. A. (1993), " Effect of Increased Productivity on the Abundances of Trophic Levels (http://www.jstor.org/stable/2462676?seq=1)", *The American Naturalist* **141** (3): 351–371,

[48] Odum, H. T. (1957). " Trophic structure and productivity of Silver Springs, Florida (http://www.esajournals.org/doi/abs/10.2307/1948571)". *Ecological Monographs* **27** (1): 55-112. .

[49] Egerton, Frank N. (2007). "Understanding Food Chains and Food Webs, 1700–1970". *Bulletin of the Ecological Society of America* **88**: 50–69. doi: 10.1890/0012-9623(2007)88[50:UFCAFW]2.0.CO;2 (http://dx.doi.org/10.1890/0012-9623(2007)88[50:UFCAFW]2.0.CO;2).

[50] Shurin, J. B.; Gruner, D. S.; Hillebrand, H. (2006), " All wet or dried up? Real differences between aquatic and terrestrial food webs. (http://rspb.royalsocietypublishing.org/content/273/1582/1.full.pdf+html)", *Proc. R. Soc. B* **273**: 1–9, doi: 10.1098/rspb.2005.3377 (http://dx.doi.org/10.1098/rspb.2005.3377),

[51] Elton, C. S. (1927). *Animal Ecology.* London, UK.: Sidgwick and Jackson.

[52] Allee, W. C. (1932). *Animal life and social growth.* Baltimore: The Williams & Wilkins Company and Associates.

[53] Edwards, J.; Fraser, K. (1983), " Concept maps as reflectors of conceptual understanding. (http://www.springerlink.com/content/64x5123271427467/)", *Research in science education* **13**: 19–26,

[54] Pimm, S. L. (2002). *Food webs.* Chicago: The University of Chicago Press.

[55] Elser, J.; Hayakawa, K.; Urabe, J. (2001). " Nutrient Limitation Reduces Food Quality for Zooplankton: Daphnia Response to Seston Phosphorus Enrichment. (http://www.esajournals.org/doi/abs/10.1890/0012-9658(2001)082[0898:NLRFQF]2.0.CO;2)". *Ecology* **82** (3): 898–903. .

[56] Worm, B.; Duffy, J.E. (2003). "Biodiversity, productivity and stability in real food webs". *Trends in Ecology and Evolution* **18** (12): 628–632. doi: 10.1016/j.tree.2003.09.003 (http://dx.doi.org/10.1016/j.tree.2003.09.003).

[57] Post, D. M. (1993). "The long and short of food-chain length". *Trends in Ecology and Evolution* **17** (6): 269–277. doi: 10.1016/S0169-5347(02)02455-2 (http://dx.doi.org/10.1016/S0169-5347(02)02455-2).

[58] Oksanen, L. (1991). "Trophic levels and trophic dynamics: A consensus emerging?". *Trends in Ecology and Evolution* **6** (2): 58–60. doi: 10.1016/0169-5347(91)90124-G (http://dx.doi.org/10.1016/0169-5347(91)90124-G).

[59] Rickleffs, Robert, E. (1996). *The Economy of Nature.* University of Chicago Press. pp. 678. ISBN 0716738473.

[60] Raffaelli, D. (2002). "From Elton to Mathematics and Back Again". *Science* **296** (5570): 1035–1037. doi: 10.1126/science.1072080 (http://dx.doi.org/10.1126/science.1072080). PMID 12004106 (http://www.ncbi.nlm.nih.gov/pubmed/12004106).

[61] Proulx, Stephen R.; Promislow, Daniel E.L.; Phillips, Patrick C. (2005). "Network thinking in ecology and evolution". *Trends in Ecology and Evolution* **20** (6): 345–353. doi: 10.1016/j.tree.2005.04.004 (http://dx.doi.org/10.1016/j.tree.2005.04.004). PMID 16701391 (http://www.ncbi.nlm.nih.gov/pubmed/16701391).

[62] Whitman, W. B.; Coleman, D. C.; Wieb, W. J. (1998). " Prokaryotes: The unseen majority (http://www.pnas.org/content/95/12/6578.full.pdf)". *Proc. Natl. Acad. Sci. USA* **95**: 6578–6583. .

[63] Groombridge, B.; Jenkins, M. (2002), *World atlas of biodiversity: earth's living resources in the 21st century* (http://books.google.ca/books?id=_kHeAXV5-XwC&printsec=frontcover&source=gbs_navlinks_s#v=onepage&q=biomass&f=false), World Conservation Monitoring Centre, United Nations Environment Programme, ISBN 0-520-23688-8,

[64] Pechmann, J. H. K. (1988). " Evolution: The Missing Ingredient in Systems Ecology (http://www.jstor.org/stable/2462267)". *The American Naturalist* **132** (9): 884–899. .

[65] Kemp, W. M. (1979). " Toward Canonical Trophic Aggregations (http://www.jstor.org/stable/2460557)". *The American Naturalist* **114** (6): 871–883. .

[66] Li, B. (2000). "Why is the holistic approach becoming so important in landscape ecology?". *Landscape and Urban Planning* **50** (1-3): 27-41. doi: 10.1016/S0169-2046(00)00078-5 (http://dx.doi.org/10.1016/S0169-2046(00)00078-5).

[67] Dunne JA, Williams RJ, Martinez ND, Wood RA, Erwin DH, Jennifer A.; Williams, Richard J.; Martinez, Neo D.; Wood, Rachel A.; Erwin, Douglas H.; Dobson, Andrew P. (2008). "Compilation and Network Analyses of Cambrian Food Webs.". *PlosBiol* **6** (4): e102. doi: 10.1371/journal.pbio.0060102 (http://dx.doi.org/10.1371/journal.pbio.0060102).

[68] Polis, G.A.; Sears, A.L.W.; Huxel, G.R.; Strong, D.R.; Maron, J. (2000). " When is a trophic cascade a trophic cascade? (http://www.cof.orst.edu/leopold/class-reading/Polis 2000.pdf)". *Trends in Ecology and Evolution* **15** (11): 473–475. doi: 10.1016/S0169-5347(00)01971-6 (http://dx.doi.org/10.1016/S0169-5347(00)01971-6). PMID 11050351 (http://www.ncbi.nlm.nih.gov/pubmed/11050351). .

[69] Mills, L.S.; Soule, M.E.; Doak, D.F. (1993). "The Keystone-Species Concept in Ecology and Conservation". *BioScience* **43** (4): 219–224. doi: 10.2307/1312122 (http://dx.doi.org/10.2307/1312122).

[70] Anderson, P.K. (1995). "Competition, predation, and the evolution and extinction of Stellar's sea cow, *Hydrodamalis gigas*". *Marine Mammal Science* **11** (3): 391–394. doi: 10.1111/j.1748-7692.1995.tb00294.x (http://dx.doi.org/10.1111/j.1748-7692.1995.tb00294.x).

[71] Igamberdiev, Abir U.; Lea, P. J. (2006). " Land plants equilibrate O_2 and CO_2 concentrations in the atmosphere. (http://www.mun.ca/biology/igamberdiev/PhotosRes_CO2review.pdf)". *Photosynthesis Research* **87** (2): 177–194. .

[72] Margulis, L. (1973). " Atmospheric homeostasis by and for the biosphere: the gaia hypothesis. (http://people.uncw.edu/borretts/courses/BIO60209/Lovelock Margulis 1974 atmospheric homeostasis by and for the biosphere - the gaia hypothesis.pdf)". *Tellus* **26**: 2–10. .

[73] Miles, D. B.; Dunham, A. E. (1993). " Historical Perspectives in Ecology and Evolutionary Biology: The Use of Phylogenetic Comparative Analyses (http://arjournals.annualreviews.org/doi/abs/10.1146/annurev.es.24.110193.003103)". *Annual Review of Ecology and Systematics* **24**: 587–619. .

[74] *Trends in Ecology and Evolution.* (http://www.cell.com/trends/ecology-evolution/home) Official Cell Press page the journal. Elsevier, Inc. 2009

[75] Vrba, E. S.; Eldredge, N. (1984), " Individuals, Hierarchies and Processes: Towards a More Complete Evolutionary Theory (http://www.jstor.org/stable/2400395)", *Paleobiology* **10** (2): 146–171,

[76] Gould, S.J.; Lloyd, E.A. (1999). "Individuality and adaptation across levels of selection: How shall we name and generalize the unit of Darwinism?". *Proceedings of the National Academy of Science* **96** (21): 11904–11909. doi: 10.1073/pnas.96.21.11904 (http://dx.doi.org/10.1073/pnas.96.21.11904).

[77] Stuart-Fox, D.; Moussalli, A. (2008). " Selection for Social Signalling Drives the Evolution of Chameleon Colour Change. (http://www. plosbiology.org/article/info:doi/10.1371/journal.pbio.0060025)". *PLoS Biol* **6** (1): e25. doi: 10.1371/journal.pbio.0060025 (http://dx. doi.org/10.1371/journal.pbio.0060025). .

[78] http://www.behavecol.com/pages/society/welcome.html

[79] Wilson, E. O. (2000). *Sociobiology: The New Synthesis* (http://books.google.ca/books?id=v7lV9tz8fXAC&printsec=frontcover& dq=sociobiology&client=firefox-a&cd=2#v=onepage&q=&f=false) (25th anniversary Ed. ed.). President and Fellows of Harvard College. ISBN 978-0674000896. .

[80] Gould, Stephen, J.; Vrba, Elizabeth, S. (1982). "Exaptation-a missing term in the science of form.". *Paleobiology* **8** (1): 4–15.

[81] Ives, A. R.; Cardinale, B. J.; Snyder, W. E. (2004), " A synthesis of subdisciplines: predator–prey interactions, and biodiversity and ecosystem functioning (http://www.lifesci.ucsb.edu/eemb/labs/cardinale/pdfs/ives_ecol_lett_2005.pdf)", *Ecology Letters* **8** (1): 102–116,

[82] Krebs, J. R.; Davies, N. B. (1993). *An Introduction to Behavioural Ecology* (http://books.google.ca/books?id=CA31asx7zq4C& printsec=frontcover&dq=behavioral+ecology+an+introduction&client=firefox-a&cd=1#v=onepage&q=&f=false). Wiley-Blackwel. pp. 432. ISBN 978-0632035465. .

[83] Webb, J. K.; Pike, D. A.; Shine, R. (2010), "Olfactory recognition of predators by nocturnal lizards: safety outweighs thermal benefits", *Behavioural Ecology* **21** (1): 72–77

[84] Cooper, W. E.; Frederick, W. G. (2010), " Predator lethality, optimal escape behavior, and autotomy (http://library.unbc.ca:3000/cgi/ content/abstract/21/1/91)", *Behavioral Ecology* **21** (1): 91–96,

[85] Fukomoto J. (1995). Long-toed salamander (*Ambystoma macrodactylum*) ecology and management in Waterton Lakes National Park. The University of Calgary, Thesis or Dissertation, M.E.Des.

[86] Toledo RC, Jared C. (1995). Cutaneous granular glands and amphibian venoms. *Comparative Biochemistry and Physiology Part A: Physiology* **111**(1):1–29. Abstract (http://www.sciencedirect.com/science?_ob=ArticleURL&_udi=B6T2P-3XWRPK4-29&_user=10& _rdoc=1&_fmt=&_orig=search&_sort=d&view=c&_acct=C000050221&_version=1&_urlVersion=0&_userid=10& md5=99d6e86855d51ed2e1d1971ae8b74223)

[87] Williams TA, Larsen JH Jr. (2005). New function for the granular skin glands of the eastern long-toed salamander, *Ambystoma macrodactylum columbianum. Journal of Experimental Zoology* **239**(3): 329–333.

[88] Grant JB, Evans JA. (2007). A technique to collect and assay adhesive-free skin secretions from *Ambystomatid* salamanders. *Herpetological Review* **38**(3):301–305.

[89] Aanena, D. K.; Hoekstra, R. F. (2007). " The evolution of obligate mutualism: if you can't beat 'em, join 'em (http://www.izb.unibe.ch/ student/Lehrveranstaltungen/pdf/07_hopa/Aanen 07.pdf)". *Trends in Ecology & Evolution* **22** (10): 506-509. .

[90] Kiers, E. T.; van der Heijden, M. G. A. (2006). " Mutualistic stability in the arbuscular mycorrhizal symbiosis: Exploring hypotheses of evolutionary cooperation. (http://people.umass.edu/lsadler/adlersite/kiers/Kiers_Ecology_2006.pdf)". *Ecology* **87** (7): 1627-1636. .

[91] Page, R. D. M. (1991). " Clocks, Clades, and Cospeciation: Comparing Rates of Evolution and Timing of Cospeciation Events in Host-Parasite Assemblages (http://www.jstor.org/pss/2992256)". *Systematic Zoology* **40** (2): 188-198. .

[92] Boucher, D. H.; James, S.; Keeler, K. H. (1982). "The Ecology of Mutualism". *Annual Review of Ecology and Systematics* **13**: 315-347.

[93] Bronstein, J. L. (1994). "Our current understanding of mutualism". *The Quarterly Review of Biology* **69** (1): 31-51.

[94] Herre, E. A.; Knowlton, N.; Mueller, U. G.; Rehner, S. A. (1999). " The evolution of mutualisms: Exploring the paths between conflict and cooperation. (http://www.biology.lsu.edu/webfac/kharms/HerreEA_etal_1999_TREE.pdf)". *Trends in Ecology and Evolution* **14** (2): 49-53. .

[95] Gilbert, F. S. (1990). *Insect life cycles: genetics, evolution, and co-ordination* (http://www.cefe.cnrs.fr/coev/pdf/fk/Addicot1990.pdf). New York: Springer-Verlag. pp. 258. .

[96] Sherman, P. W.; Lacey, E. A.; Reeve, H. K.; Keller, L. (1995). " The eusociality continuum (http://www.nbb.cornell.edu/neurobio/ BioNB427/READINGS/ShermanEtAl1995.pdf)". *Behavioural Ecology* **6** (1): 102-108. .

[97] Strassmann, J. E. (2000). " Altruism and social cheating in the social amoeba Dictyostelium discoideum (http://www.nature.com/nature/ journal/v408/n6815/abs/408965a0.html)". *Nature* **408**: 965-967. doi: 10.1038/35050087 (http://dx.doi.org/10.1038/35050087). .

[98] Wilson, D. S.; Wilson, E. O. (2007). " Rethinking the theoretical foundation of sociobiology (http://evolution.binghamton.edu/dswilson/ resources/publications_resources/Rethinking sociobiology.pdf)". *The Quarterly Review of Biology, December 2007, Vol. 82, No. 4* **82** (4): 327-348. .

[99] Amundsen, T.; Slagsvold, T. (1996), "Lack's Brood Reduction Hypothesis and Avian Hatching Asynchrony: What's Next?", *Oikos* **76** (3): 613–620

[100] Pijanowski, B. C. (1992), "A Revision of Lack's Brood Reduction Hypothesis", *The American Naturalist* **139** (6): 1270–1292

[101] Lack, D. (1956), " Variations in the Reproductive Rate of Birds (http://www.jstor.org/stable/82998)", *Proceedings of the Royal Society of London. Series B, Biological Sciences* **145** (920): 329–333,

[102] Kodric-Brown, A.; Brown, J. H. (1984), " Truth in advertising: The kinds of traits favored by sexual selection (http://dbs.umt.edu/ courses/biol406/readings/Wk6-Kodric-Brown and Brown 1984.pdf)", *The American Naturalist* **124** (3): 309–323,

[103] Parenti, L. R.; Ebach, M. C. (2009), *Comparative biogeography: Discovering and classifying biogeographical patterns of a dynamic Earth.* (http://books.google.ca/books?id=K1GU_l16bG4C&printsec=frontcover&source=gbs_v2_summary_r&cad=0#v=onepage&q=& f=false), London, England: University of California Press, ISBN 978-0-520-25945-4,

[104] http://www.wiley.com/bw/journal.asp?ref=0305-0270

[105] Wiens, J. J.; Donoghue, M. J. (2004), " Historical biogeography, ecology and species richness (http://www.phylodiversity.net/ donoghue/publications/MJD_papers/2004/144_Wiens_TREE04.pdf)", *Trends in Ecology and Evolution* **19** (12): 639–644,

[106] Croizat, L.; Nelson, G.; Rosen, D. E. (1974), " Centers of Origin and Related Concepts (http://www.jstor.org/stable/2412139)", *Systematic Zoology* **23** (2): 265–287,

[107] Wiley, E. O. (1988), " Vicariance Biogeography (http://www.jstor.org/stable/2097164)", *Annual Review of Ecology and Systematics* **19**: 513–542,

[108] Morrone, J. J.; Crisci, J. V. (1995), " Historical Biogeography: Introduction to Methods (http://arjournals.annualreviews.org/doi/abs/ 10.1146/annurev.es.26.110195.002105)", *Annual Review of Ecology and Systematics* **26**: 373–401,

[109] Landhäusser, Simon M.; Deshaies, D.; Lieffers, V. J. (2009), " Disturbance facilitates rapid range expansion of aspen into higher elevations of the Rocky Mountains under a warming climate (http://www3.interscience.wiley.com/journal/122574329/abstract)", *Journal of Biogeography* **37** (1): 68–76,

[110] Svenning, Jens-Christian; Condi, R. (2008), " Biodiversity in a Warmer World (http://www.sciencemag.org/cgi/content/full/322/ 5899/206)", *Science* **322** (5899): 206–207,

[111] Santos, J. C.; Coloma, L. A.; Summers, K.; Caldwell, J. P.; Ree, R.; et al. (2009). " Amazonian Amphibian Diversity Is Primarily Derived from Late Miocene Andean Uplift. (http://www.plosbiology.org/article/info:doi/10.1371/journal.pbio.1000056)". *PLoS Biol* **7** (3): e1000056. doi: 10.1371/journal.pbio.1000056 (http://dx.doi.org/10.1371/journal.pbio.1000056). .

[112] http://www3.interscience.wiley.com/journal/117989598/home]

[113] Avise, J. (1994). *Molecular Markers, Natural History and Evolution* (http://books.google.ca/books?id=2zYnQfnXNr8C& printsec=frontcover&dq=john+avise+molecular&client=firefox-a&cd=1#v=onepage&q=&f=false). Kluwer Academic Publishers. ISBN 0-412-03771-8. .

[114] Patton, J. L., and Smith, M. F. (1993). Molecular evidence for mating asymmetry and female choice in a pocket gopher (*Thomomys*) hybrid zone. Molecular Ecology, 2, 3-8.

[115] O'Brian, E.; Dawson, R. (2007). " Context-dependent genetic benefits of extra-pair mate choice in a socially monogamous passerine (http:/ /web.unbc.ca/~dawsonr/2007_bes61_775-782.pdf)". *Behav Ecol Sociobiol* **61**: 775–782. doi: 10.1007/s00265-006-0308-8 (http://dx.doi. org/10.1007/s00265-006-0308-8). .

[116] Avise, J. (2000). *Phylogeography: The History and Formation of Species* (http://books.google.ca/books?id=lA7YWH4M8FUC& printsec=frontcover&dq=phylogeography&client=firefox-a&cd=1#v=onepage&q=&f=false). President and Fellows of Harvard College. ISBN 0-674-66638-0. .

[117] http://zipcodezoo.com/Glossary/holocoenotic.asp

[118] Campbell, Neil A.; Brad Williamson; Robin J. Heyden (2006). *Biology: Exploring Life* (http://www.phschool.com/el_marketing.html). Boston, Massachusetts: Pearson Prentice Hall. ISBN 0-13-250882-6. .

[119] Kormondy, E. (1995). *Concepts of ecology. (4th ed.)*. Benjamin Cummings. ISBN 0134781163.

[120] Ernst, S. K. Morgan; Enquist, Brian J.; Brown, James H.; Charnov, E. L.; Gillooly, J. F.; Savage, Van M.; et al. (2003). " Thermodynamic and metabolic effects on the scaling of production and population energy use (https://www.msu.edu/~maurerb/Ernest_etal_2003.pdf)". *Ecology Letters* **6**: 990–995. doi: 10.1046/j.1461-0248.2003.00526.x (http://dx.doi.org/10.1046/j.1461-0248.2003.00526.x). .

[121] Allègre, Claude J.; Manhès, Gérard; Göpel, Christa (1995). " The age of the Earth (http://www.sciencedirect.com/ science?_ob=ArticleURL&_udi=B6V66-3YYTKC0-7Y&_user=10&_rdoc=1&_fmt=&_orig=search&_sort=d&_docanchor=&view=c& _searchStrId=1001748320&_rerunOrigin=scholar.google&_acct=C000050221&_version=1&_urlVersion=0&_userid=10& md5=c2e364efb25d1f6a73686ae3e7701b26)". *Geochimica et Cosmochimica Acta* **59**: 1455–1456. .

[122] Wills, C.; Bada, J. (2001). *The Spark of Life: Darwin and the Primeval Soup* (http://books.google.ca/books?id=UrGqxy0wMdkC& dq=The+Spark+of+Life:+Darwin+and+the+Primeval+Soup&printsec=frontcover&source=bl&ots=cpuX3xktry& sig=2ySEa55w1ca6yXXZcEf_fJovq_4&hl=en&ei=F7miSpqBAo6uswO3v4CNDw&sa=X&oi=book_result&ct=result& resnum=1#v=onepage&q=&f=false). Cambridge, Massachusetts: Perseus Publishing. .

[123] Catling, D. C.; Claire, M. W. (2005). " How Earth's atmosphere evolved to an oxic state: A status report (http://www.atmos.washington. edu/~davidc/papers_mine/Catling2005-EPSL.pdf)". *Earth and Planetary Science Letters* **237**: 1–20. doi: 10.1016/j.epsl.2005.06.013 (http:/ /dx.doi.org/10.1016/j.epsl.2005.06.013). .

[124] Cronk, J. K.; Fennessy, M. S. (2001), *Wetland Plants: Biology and Ecology* (http://books.google.ca/books?id=FNl1GFbH2eQC& printsec=frontcover&dq=wetland+plants&client=firefox-a&cd=1#v=onepage&q=&f=false), Washington, D.C.: Lewis Publishers, ISBN 1-56670-372-7,

[125] Evans, D. H.; Piermarini, P. M.; Potts, W. T. W. (1999), " Ionic Transport in the Fish Gill Epithelium (http://people.biology.ufl.edu/ devans/DHEJEZ.pdf)", *Journal of Experimental Zoology* **283**: 641–652,

[126] Wheeler, T. D.; Stroock, A. D. (2008). " The transpiration of water at negative pressures in a synthetic tree (http://www.nature.com/ nature/journal/v455/n7210/abs/nature07226.html)". *Nature* **455**: 208–212. .

[127] Pockman, W. T.; Sperry, J. S.; O'Leary, J. W. (1995). " Sustained and significant negative water pressure in xylem (http://www.nature. com/nature/journal/v378/n6558/abs/378715a0.html)". *Nature* **378**: 715–716. .

[128] Kastak, D.; Schusterman, R. J. (1998), " Low-frequency amphibious hearing in pinnipeds: Methods, measurements, noise, and ecology (http://www.sea-inc.net/resources/lrni_KastakandSchusterman_JASA_LFpinnipedhearing_1998.pdf)", *J. Acoust. Soc. Am.* **103** (4): 2216–2228,

[129] Shimeta, J.; Jumars, P. A.; Lessard, E. J. (1995). " Influences of turbulence on suspension feeding by planktonic protozoa; experiments in laminar shear fields (http://www.aslo.org/lo/toc/vol_40/issue_5/0845.pdf)". *Limnolology and Oceanography* **40** (5): 845–859. .

[130] Etemad-Shahidi, A.; Imberger, J. (2001). " Anatomy of turbulence in thermally stratified lakes (http://nospam.aslo.org/lo/toc/vol_46/issue_5/1158.pdf)". *Limnolology and Oceanography* **46** (5): 1158–1170. .

[131] Wolf, B. O.; Walsberg, G. E. (2006), " Thermal Effects of Radiation and Wind on a Small Bird and Implications for Microsite Selection (http://www.jstor.org/stable/2265716)", *Ecology* **77** (7): 2228–236,

[132] Lenton, T. M.; Watson, A. (2000), " Redfield revisited 2. What regulates the oxygen content of the atmosphere. (http://lgmacweb.env.uea.ac.uk/esmg/papers/Redfield_revisited_2.pdf)", *Global biogeochemical cycles* **14** (1): 249–268,

[133] Lobert, J. M.; Warnatz, J. (1993), Crutzen, P. J.; Goldammer, J. G., eds., *Emissions from the combustion process in vegetation.* (http://jurgenlobert.org/papers_data/Lobert.Warnatz.Wiley.1993.pdf), John Wiley & Sons, ISBN 0471936049, 9780471936046,

[134] Garren, K. H. (1943), " Effects of Fire on Vegetation of the Southeastern United States (http://www.springerlink.com/content/a70310371q6l1414/)", *Botanical Review* **9** (9): 617–654,

[135] Cooper, C. F. (1960), " Changes in Vegetation, Structure, and Growth of Southwestern Pine Forests since White Settlement (http://www.jstor.org/stable/1948549)", *Ecological Monographs* **30** (2): 130–164,

[136] Cooper, C. F. (1961), "The ecology of fire", *Scientific American* **204**: 150–160

[137] http://www.fireecology.net/

[138] van Wagtendonk, Jan W. (2007), " History and Evolution of Wildland Fire Use (http://fireecology.net/Journal/pdf/Volume03/Issue02/003.pdf)", *Fire Ecology Special Issue* **3** (2): 3–17,

[139] Boerner, R. E. J. (1982), " Fire and Nutrient Cycling in Temperate Ecosystems (http://www.jstor.org/stable/1308941)", *BioScience* **32** (3): 187–192,

[140] Goubitz, S.; Werger, M. J. A.; Ne'eman, G. (2009), " Germination Response to Fire-Related Factors of Seeds from Non-Serotinous and Serotinous Cones (http://www.springerlink.com/content/w28p341482tj4g4w/)", *Plant Ecology* **169** (2): 195–204,

[141] Ne'eman, G.; Goubitz, S.; Nathan, R. (2004), "Reproductive Traits of Pinus halepensis in the Light of Fire: A Critical Review", *Plant Ecology* **171** (1/2): 69–79

[142] Flematti, Gavin R.; Ghisalberti, Emilio L.; Dixon, Kingsley W.; Trengove, R. D. (2004), " A Compound from Smoke That Promotes Seed Germination (http://www.ice.mpg.de/main/news/positions/itb-004/DixonSmokepaper.pdf)", *Science* **305. no. 5686, p. 977** (5686): 977,

[143] Falkowski, P. G.; Fenchel, T.; Delong, E. F. (2008). " The microbial engines that drive Earth's biogeochemical cycles (http://www.sciencemag.org/cgi/reprint/320/5879/1034.pdf)". *Science* **320**. .

[144] Grace, J. (2004). "Understanding and managing the global carbon cycle". *Journal of Ecology* **92**: 189–202. doi:10.1111/j.0022-0477.2004.00874.x (http://dx.doi.org/10.1111/j.0022-0477.2004.00874.x).

[145] Pearson, P. N.; Palmer, M. R. (2000), " Atmospheric carbon dioxide concentrations over the past 60 million years (http://paleolands.com/pdf/cenozoicCO2.pdf)", *Nature* **406**: 695–699,

[146] Pagani, M.; Zachos, J. C.; Freeman, K. H.; Tipple, B.; Bohaty, S. (2005), " Marked Decline in Atmospheric Carbon Dioxide Concentrations During the Paleogene (http://earth.geology.yale.edu/~mp364data/Pagani.Science.2005.pdf)", *Science* **309**: 600–603,

[147] Cox, P. M.; Betts, R. A.; Jones, C. D.; Spall, S. A.; Totterdell, I. J. (2000), " Acceleration of global warming due to carbon-cycle feedbacks in a coupled climatemodel (https://www.up.ethz.ch/education/biogeochem_cycles/reading_list/cox_etal_nat_00.pdf)", *Nature* **408**: 184–187,

[148] Heimann, Martin; Reichstein, Markus (2008), " Terrestrial ecosystem carbon dynamics and climate feedbacks (http://courses.washington.edu/ocean450/Discussion_Topics_Papers/Heimmann_clim_chng_08.pdf)", *Nature* **451**: 289–292,

[149] Davidson, Eric A.; Janssens, Ivan A. (2006), " Temperature sensitivity of soil carbon decomposition and feedbacks to climate change (http://whrc.org/resources/published_literature/pdf/DavidsonetalNature.06.pdf)", *Nature* **440**: 165–173,

[150] Zhuan, Q.; Melillo, J. M.; McGuire, A. D.; Kicklighter, D. W.; Prinn, R. G.; Steudler, P. A. (2007), " Net emission of CH_4 and CO_2 in Alaska: Implications for the region's greenhouse gas budget. (http://picea.sel.uaf.edu/manuscripts/zhuang07-ea.pdf)", *Ecological Applications* **17** (1): 203–212,

[151] Kingsland, S. (2004), " Conveying the intellectual challenge of ecology: an historical perspective (http://www.isa.utl.pt/dbeb/ensino/txtapoio/HistEcology.pdf)", *Front Ecol Environ* **2** (7): 367–374,

[152] Egerton, F. N. (2001). " A History of the Ecological Sciences: Early Greek Origins (http://www.jstor.org/stable/20168519?seq=1)". *Bulletin of the Ecological Society of America* **82** (1): 93–97. .

[153] Keller, D. R.; Golley, F. B. (2000), *The philosophy of ecology: from science to synthesis.* (http://books.google.ca/books?id=uYOxUAJThJEC&pg=PP1&dq=The+philosophy+of+ecology:+from+science+to+synthesis.&client=firefox-a&cd=1#v=onepage&q=&f=false), Athens, GA: University of Georgia Press, ISBN 978-0820322209,

[154] Real, L. A.; Brown, J. H. (1992), *Foundations of ecology: classic papers with commentaries.* (http://books.google.ca/books?id=y2wwTZgrHmYC&dq=Foundations+of+ecology:+classic+papers+with+commentaries.&client=firefox-a&cd=1), Chicago: University of Chicago Press, ISBN 978-0226705941,

[155] Esbjorn-Hargens, S. (2005). " Integral Ecology: An Ecology of Perspectives (http://www.vancouver.wsu.edu/fac/tissot/IU_Ecology_Intro.pdf)". *Journal of Integral Theory and Practice* **1** (1): 2–37. .

[156] Hinchman, L. P.; Hinchman, S. K. (2007), " What we owe the Romantics (http://www.ingentaconnect.com/content/whp/ev/2007/00000016/00000003/art00006)", *Environmental Values* **16** (3): 333-354,

[157] Hughes, J. D. (1985), " Theophrastus as Ecologist (http://www.jstor.org/pss/3984460)", *Environmental Review* **9** (4): 296-306,

[158] Goodland, R. J. (1975), " The Tropical Origin of Ecology: Eugen Warming's Jubilee (http://www.jstor.org/pss/3543715)", *Oikos* **26** (2): 240-245,

[159] Kormandy, E. J. (1978). " Review: Ecology/Economy of Nature--Synonyms? (http://www.jstor.org/pss/1938247)". *Ecology* **59** (6): 1292–1294. .

[160] Stauffer, R. C. (1957), " Haeckel, Darwin and ecology. (http://www.clt.astate.edu/aromero/ECO3.Haeckel.pdf)", *The Quarterly Review of Biology* **32** (2): 138–144,

[161] Rosenzweig, M.L. (2003). " Reconciliation ecology and the future of species diversity (http://eebweb.arizona.edu/COURSES/Ecol302/Lectures/ORYXRosenzweig.pdf)". *Oryx* **37** (2): 194–205. .

[162] Hawkins, B. A. (2001). " Ecology's oldest pattern. (http://www4.ncsu.edu/~rrdunn/Hawkins 2001.pdf)". *Endeavor* **25** (3): 133. .

[163] Hughes, J. D. (1975). " Ecology in ancient Greece (http://www.informaworld.com/smpp/content~content=a902027058&db=all)". *Inquiry* **18** (2): 115–125. .

[164] Benson, Keith R. (2000). " The emergence of ecology from natural history (http://www.sciencedirect.com/science?_ob=ArticleURL& _udi=B6V81-414X355-V&_user=1067466&_rdoc=1&_fmt=&_orig=search&_sort=d&_docanchor=&view=c& _searchStrId=1143381236&_rerunOrigin=scholar.google&_acct=C000051249&_version=1&_urlVersion=0&_userid=1067466& md5=07093484296081185c20fff99e870aab)". *Endeavour* **24** (2): 59–62. .

[165] Forbes, S. (1887). " The lake as a microcosm (http://www.uam.es/personal_pdi/ciencias/scasado/documentos/Forbes.PDF)". *Bull. of the Scientific Association* (Peoria, IL : .): 77–87. .

[166] Darwin, Charles (1859). *On the Origin of Species* (http://darwin-online.org.uk/content/frameset?itemID=F373&viewtype=text& pageseq=16) (1st ed.). London: John Murray. p. 1. .

[167] Meysman, F. J. R.; Middelburg, Jack J.; Heip, C. H. R. (2006), " Bioturbation: a fresh look at Darwin's last idea (http://www.marbee. fmns.rug.nl/pdf/marbee/2006-Meysman-TREE.pdf)", *TRENDS in Ecology and Evolution* **21** (22): 688–695,

[168] Hector, A.; Hooper, R. (2002). "Darwin and the First Ecological Experiment". *Science* **295**: 639–640.

[169] Acot, P. (1997). "The Lamarckian Cradle of Scientific Ecology". *Acta Biotheoretica* **45** (3-4): 185–193.

[170] Clements, F. E. (1905). *Research Methods in Ecology.* Lincoln, Nebraska: University Publ..

[171] Simberloff, D. (1980). "A succession of paradigms in ecology: Essentialism to materialism and probalism.". *Synthese* **43** (**1980**) 3-39: 3–39.

[172] Gleason, H. A. (1926). " The Individualistic Concept of the Plant Association (http://www.ecologia.unam.mx/laboratorios/ comunidades/pdf/pdf curso posgrado Elena/Tema 1/gleason1926.pdf)". *Bulletin of the Torrey Botanical Club* **53** (1): 7–26. .

[173] Weltzin, J. F.; Belote, R. T.; Williams, L. T.; Engel, E. C. (2006). " Authorship in ecology: attribution, accountability, and responsibility (http://www.biology.duke.edu/jackson/ecophys/WeltzinFrontiers2006.pdf)". *Frontiers in Ecology and the Environment* **4** (8): 435-441. .

[174] McIntosh, R. P. (1989). " Citation Classics of Ecology (http://www.jstor.org/pss/2831684)". *The Quarterly Review of Biology* **64** (1): 31-49. .

[175] Leimu, R.; Koricheva, J. (2005). " What determines the citation frequency of ecological papers? (http://www.mbt.mtesz.hu/ 081okologiai_szakosztaly/citations in ecology.pdf)". *Trends in Ecology & Evolution* **20** (1): 28-32. .

[176] Damschen, E. I.; Rosenfeld, K. M.; Haddad, N. M. (2005), " Visibility Matters: Increasing Knowledge of Women's Contributions to Ecology (http://biology4.wustl.edu/faculty/damschen/Pubs/Damschen et al.2005.pdf)", *Frontiers in Ecology and the Enivornment* **3** (4): 212-219,

[177] Ghilarov, A. M. (1995). " Vernadsky's Biosphere Concept: An Historical Perspective (http://www.jstor.org/pss/3036242)". *The Quarterly Review of Biology* **70** (2): 193-203. .

[178] Itô, Y. (1991). " Development of ecology in Japan, with special reference to the role of Kinji Imanishi (http://www.springerlink.com/ content/64856221n5746428/)". *Journal of Ecological Research* **6** (2): 139-155. .

[179] http://www4.ncsu.edu/~rrdunn/Hawkins%202001.pdf

[180] Futuyma, D. J. (2005). The Nature of Natural Selection. Ch. 8, pages 93-98 in Cracraft, J. and Bybee R. W. (Eds.) Evolutionary Science and Society: Educating a New Generation. American Institute of Biological Sciences.

[181] Glaubrecht, M. (2008), " Homage to Karl August Möbius (1825-1908) and his contributions to biology: zoologist, ecologist, and director at the Museum für Naturkunde in Berlin (http://www3.interscience.wiley.com/journal/117944310/abstract)", *Zoosystematics and Evolution* **84** (1): 9–30,

[182] Baker, H. G. (1966). " Resoning about adaptations in ecosystems (http://www.jstor.org/pss/1293551)". *BioScience* **16** (1): 35–37. .

[183] Nyhart, L. K. (1998). " Civic and Economic Zoology in Nineteenth-Century Germany: The "Living Communities" of Karl Mobius (http:// www.jstor.org/pss/236735)". *Isis* **89** (4): 605–630. .

[184] Coleman, W. (1986). " Evolution into ecology? The strategy of warming's ecological plant geography (http://www.springerlink.com/ content/l620506185326641/)". *Journal of the History of Biology* **19** (2): 181–196. .

[185] Palamar, C. R. (2008), " The Justice of Ecological Restoration: Environmental History, Health, Ecology, and Justice in the United States (http://www.humanecologyreview.org/pastissues/her151/palamar.pdf)", *Human Ecology Review* **15** (1): 82-94,

[186] http://www.uam.es/personal_pdi/ciencias/scasado/documentos/Forbes.PDF

[187] Forbes, S. A. (1915). " The ecological foundations of applied entomology (http://www.uam.es/personal_pdi/ciencias/scasado/ documentos/Forbes.PDF)". *Annals of the Entomological Society of America* **8** (1): 1–19. .

[188] Adams, C. C.; Fuller, G. D. (1940), " Henry Chandler Cowles, Physiographic Plant Ecologist (http://www.jstor.org/stable/2561130)", *Annals of the Association of American Geographers* **31** (1): 39–43,

[189] Cooper, W. S. (1957), " Sir Arthur Tansley and the Science of Ecology (http://www.jstor.org/stable/1943136)", *Ecology* **38** (4): 658–659,

[190] Kingsland, S. E. (1994). " Review: The History of Ecology (http://www.springerlink.com/content/m51188130814962k/)". *Journal of the History of Biology* **27** (2): 349–357. .

[191] Ilerbaig, J. (1999), " Allied Sciences and Fundamental Problems: C.C. Adams and the Search for Method in Early American Ecology (http://www.jstor.org/stable/4331545?&Search=yes&term=Adams&term=contributions&term=C.&list=hide&searchUri=/action/doBasicSearch?Query=C.+C.+Adams+contributions&wc=on&dc=All+Disciplines&item=1&ttl=41072& returnArticleService=showArticle)", *Journal of the History of Biology* **32**: 439–463,

[192] Raup, H. M. (1959), " Charles C. Adams, 1873-1955 (http://www.jstor.org/stable/2561526?&Search=yes&term=contributions& term="Charles+C.+Adams"&list=hide&searchUri=/action/doBasicSearch?Query=%22Charles+C.+Adams%22+contributions& gw=jtx&prq=Charles+C.+Adams+contributions&Search=Search&hp=25&wc=on&item=1&ttl=113& returnArticleService=showArticle)", *Annals of the Association of American Geographers* **49** (2): 164–167,

[193] Ellison, A. M. (2006), " What Makes an Ecological Icon. (http://www.esajournals.org/doi/abs/10.1890/ 0012-9623(2006)87[380:WMAEI]2.0.CO;2)", *Bulletin of the Ecological Society of America* **87** (4): 380–386,

[194] Kendeigh, S. C. (1968), " Victor Ernest Shelford, Eminent Ecologist, 1968 (http://www.jstor.org/stable/20165761?&Search=yes& term="Victor+Ernest+Shelford"&term=contributions&list=hide&searchUri=/action/doBasicSearch?Query=%22Victor+Ernest+ Shelford%22+contributions&gw=jtx&prq=%22Karl+M%C3%B6bius%22+contributions&Search=Search&hp=25&wc=on&item=5& ttl=33&returnArticleService=showArticle)", *Bulletin of the Ecological Society of America* **49** (3): 97–100,

[195] McIntosh, R. P. (1975), " H. A. Gleason-"Individualistic Ecologist" 1882-1975: His Contributions to Ecological Theory (http://www. jstor.org/stable/2484142)", *1975* **105** (5): 253–278,

[196] Southwood, R.; Clarke, J. R. (1999), " Charles Sutherland Elton. 29 March 1900-1 May 1991 (http://www.jstor.org/stable/ 770268?seq=1&Search=yes&term="Charles+Elton"&term=contributions&list=hide&searchUri=/action/doBasicResults?hp=25&la=& wc=on&gw=jtx&jcpsi=1&artsi=1&Query=%22Charles+Elton%22+contributions&sbq=%22Charles+Elton%22+contributions& prq=%22Charles+C.+Adams%22+contributions&si=26&jtxsi=26&ttl=29&returnArticleService=showArticle& resultsServiceName=doBasicResultsFromArticle)", *Biographical Memoirs of Fellows of the Royal Society* **45**: 131–146,

[197] Flannery, M. C. (2003), " Evelyn Hutchinson: A Wonderful Mind (http://www.jstor.org/stable/4451536?&Search=yes& term="Evelyn+Hutchinson"&term=contributions&list=hide&searchUri=/action/doBasicSearch?Query=%22Evelyn+Hutchinson%22+ contributions&wc=on&dc=All+Disciplines&item=2&ttl=273&returnArticleService=showArticle)", *The American Biology Teacher* **65** (6): 462–467,

[198] Edmondson, Y. H. (1991), " In Memoriam: G. Evelyn Hutchinson, 1903-1991 (http://www.jstor.org/stable/2837527)", *Limnology and Oceanography* **36** (3): 617,

[199] Patrick, R. (1994), " George Evelyn Hutchinson (30 January 1903-17 May 1991) (http://www.jstor.org/stable/986851)", *Proceedings of the American Philosophical Society* **138** (4): 531–535,

[200] Gunderson, L.; Folke, C.; Lee, M.; Holling, C. S. (2002), " In memory of mavericks. (http://www.consecol.org/vol6/iss2/art19/)", *Conservation Ecology* **6** (2): 19,

[201] Rotabi, K. S. (2007), " Ecological Theory Origin from Natural to Social Science of Vice Versa? A Brief Conceptual History for Social Work (http://journals.iupui.edu/index.php/advancesinsocialwork/article/viewFile/135/136)", *Advances in Social Work* **8** (1): 113–129,

[202] Patten, B. C. (1993), " Toward a more holistic ecology, and science: the contribution of H.T. Odum (http://www.springerlink.com/ content/h831866562302236/)", *Oecologia* **93** (4): 597–602,

[203] Ewel, J. J. (2003), " Howard Thomas Odum (1924–2002) (http://www.esajournals.org/doi/abs/10.1890/ 0012-9623(2003)84[13:HTO]2.0.CO;2)", *Bulletin of the Ecological Society of America* **84** (1): 13–15,

[204] Brown, J. H. (1999), " The Legacy of Robert Macarthur: From Geographical Ecology to Macroecology (http://www.jstor.org/pss/ 1383283)", *Journal of Mammalogy* **80** (2): 333-344,

[205] Becker, C. D.; Ostrom, E. (1995). " Human Ecology and Resource Sustainability: The Importance of Institutional Diversity (http://www. umich.edu/~ifri/Publications/R951_20.pdf)". *Annual Review of Ecology and Systematics* **26**: 113-133. doi: 10.1146/annurev.es.26.110195.000553 (http://dx.doi.org/10.1146/annurev.es.26.110195.000553). .

[206] Hartig, T. (2008). " Green space, psychological restoration, and health inequality (http://www.sciencedirect.com/ science?_ob=ArticleURL&_udi=B6T1B-4TVTJ12-7&_user=10&_rdoc=1&_fmt=&_orig=search&_sort=d&_docanchor=&view=c& _searchStrId=1149870024&_rerunOrigin=scholar.google&_acct=C000050221&_version=1&_urlVersion=0&_userid=10& md5=bd504dbea84447e7297be383f977e01d)". *The Lancet* **372** (9650): 1614-1615. .

[207] Pickett, S. T. A.; Cadenasso, M. L. (2007). " Linking ecological and built components of urban mosaics: an open cycle of ecological design (http://www.ecostudies.org/pickett/2008_Ecological_Built_J_Ecol.pdf)". *Journal of Ecology* **96**: 8-12. .

[208] Termorshuizen, J. W.; Opdam, P.; van den Brink, A. (2007). " Incorporating ecological sustainability into landscape planning (http:// www.ontwerpenmetnatuur.wur.nl/NR/rdonlyres/EBE08632-E0F4-4DA1-ACDE-5C4C0710056C/44955/termorshuizenetal.pdf)". *Landscape and Urban Planning* **79** (3-4): 374-384. .

[209] Díaz, S.; Fargione, J.; Chapin, F. S.; Tilman, D. (2006). " Biodiversity Loss Threatens Human Well-Being. (http://www.plosbiology.org/ article/info:doi/10.1371/journal.pbio.0040277)". *PLoS Biol* **4** (8): e277. doi: doi:10.1371/journal.pbio.0040277 (http://dx.doi.org/doi:10. 1371/journal.pbio.0040277). .

[210] Ostrom, E.; Burger, J.; Field, C. B.; Norgaard, R. B.; Policansky, D. (1999). " Revisiting the Commons: Local Lessons, Global Challenges (http://isites.harvard.edu/fs/docs/icb.topic464862.files/Revisiting_the_Commons.pdf)". *Science* **284**: 278-28. .

[211] http://www.millenniumassessment.org/en/Synthesis.aspx

[212] Sienkiewicz, A. (2006). " Toward a Legal Land Ethic: Punitive Damages, Natural Value, and the Ecological Commons (https:// litigation-essentials.lexisnexis.com/webcd/app?action=DocumentDisplay&crawlid=1&doctype=cite&docid=15+Penn+St.+Envtl.+L. +Rev.+91&srctype=smi&srcid=3B15&key=6d0993165c3d310fcc3ceb54672154db)". *Penn State Environmental Law Review* **91**: 95-96. .

[213] de Groot, R. S.; Wilson, M. A.; Boumans, R. M. J. (2002). " A typology for the classification, description and valuation of ecosystem functions, goods and services (http://yosemite.epa.gov/SAB/sabcvpess.nsf/e1853c0b6014d36585256dbf005c5b71/ 1c7c986c372fa8d485256e29004c7084/$FILE/deGroot et al.pdf)". *Ecological Economics* **41**: 393-408. .

[214] http://www.elsevier.com/wps/find/journaldescription.cws_home/503305/description#description

[215] http://www.ecoeco.org/

[216] {Cite journal I last = Costanza I first = R. I last2 = d'Arge I first2 = R. I last3 = de Groot I first3 = R. I last4 = Farberk I first4 = S. I last5 = Grasso I first5 = M. I last6 = Hannon I first6 = B. I last = et al. I title = The value of the world's ecosystem services and natural capital. I journal = Nature I volume = 387 I pages = 253-260 I year = 1997 I url = http://www.uvm.edu/giee/publications/Nature_Paper.pdf}}

[217] Ceballos, G.; Ehrlich, P. R. (2002). " Mammal Population Losses and the Extinction Crisis (http://epswww.unm.edu/facstaff/gmeyer/ envsc330/CeballosEhrlichmammalextinct2002.pdf)". *Science* **296** (5569): 904-907. .

[218] Rees, W. E. (1992). " Ecological footprints and appropriated carrying capacity: what urban economics leaves out. (http://eau.sagepub. com/cgi/reprint/4/2/121)". *Environment and Urbanization* **4** (2): 121-130. .

[219] http://assets.panda.org/downloads/living_planet_report_2008.pdf

[220] Moran, D. D.; Wakernagel, M.; Kitzesa, J. A.; Goldfinger, S. H.; Boutau, A. (2008). " Measuring sustainable development — Nation by nation (http://www.rshanthini.com/tmp/CP551/M02R02_MeasuringSDwithHDIandEF.pdf)". *Ecological Economics* **64**: 470-474. .

[221] Rees, W. (2002). " An Ecological Economics Perspective on Sustainability and Prospects for Ending Poverty (http://www.springerlink. com/content/g20265734n8670q8/)". *Population & Environment* **24** (1): 15-46. .

[222] Dasgupta, P. (2008). " Creative Accounting (http://www.nature.com/nature/journal/v456/n1s/full/twas08.44a.html)". *Nature Frontiers* **456**: 44. doi:10.1038/twas08.44a (http://dx.doi.org/10.1038/twas08.44a). .

[223] Wackernagel, M.; Rees, W. E. (1997). " Perceptual and structural barriers to investing in natural capital: Economics from an ecological footprint perspective (http://www.sciencedirect.com/science?_ob=ArticleURL&_udi=B6VDY-3SVHN6G-2&_user=10&_rdoc=1& _fmt=&_orig=search&_sort=d&_docanchor=&view=c&_acct=C000050221&_version=1&_urlVersion=0&_userid=10& md5=f6e4fea3c3c369ae70540daf1f8b92ff)". *Ecological Economics* **20** (1): 3-24. doi: 10.1016/S0921-8009(96)00077-8 (http://dx.doi.org/ 10.1016/S0921-8009(96)00077-8). .

[224] Pastor, J.; Light, S.; Sovel, L. (1998). " Sustainability and resilience in boreal regions: sources and consequences of variability. (http:// www.consecol.org/vol2/iss2/art16/)". *Conservation Ecology* **2** (2): 16. .

[225] Zalasiewicz, J.; Williams, M.; Alan, S.; Barry, T. L.; Coe, A. L.; Bown, P. R.; et al. (2008). " Are we now living in the Anthropocene (http://www.see.ed.ac.uk/~shs/Climate change/Geo-politics/Anthropocene 2.pdf)". *GSA Today* **18** (2): 4-8. .

[226] Wake, D. B.; Vredenburg, V. T. (2008). " Are we in the midst of the sixth mass extinction? A view from the world of amphibians (http:// www.pnas.org/content/105/suppl.1/11466.full)". *PNAS* **105**: 11466-11473. doi: 10.1073/pnas.0801921105 (http://dx.doi.org/10.1073/ pnas.0801921105). .

[227] Mooney, H.; Larigauderie, A.; Cesario, M.; Elmquist, T.; Hoegh-Guldberg, O.; Lavorel, S.; et al. (2009). " Biodiversity, climate change, and ecosystem services Current Opinion in Environmental Sustainability (http://www.sciencedirect.com/science?_ob=ArticleURL& _udi=B985C-4WY5BTH-1&_user=1067466&_rdoc=1&_fmt=&_orig=search&_sort=d&_docanchor=&view=c&_acct=C000051249& _version=1&_urlVersion=0&_userid=1067466&md5=7586a0d8a93b391b9fcb00d1b34881d4)". *Current Opinion in Environmental Sustainability* **1** (1): 46-54. .

[228] Chapin, F. S.; Zaveleta, E. S.; Eviner, V. T.; Naylor, R. L.; Vitousek, P. M.; Reynolds, H. L.; et al. (2000). " Consequences of changing biodiversity (http://dx.doi.org/10.1038/35012241)". *Nature* **405** (6783): 234-242. .

[229] Ehrlich, P. R.; Pringle, R. M. (2008). " Where does biodiversity go from here? A grim business-as-usual forecast and a hopeful portfolio of partial solutions. (http://www.pnas.org/content/105/suppl.1/11579.full)". *Proceedings of the National Academy of Sciences* **105** (S1): 11579-11586. .

[230] Koh, L. P.; Dunn, R. R.; Sodhi, N. S.; Colwell, R. K.; Proctor, H. C.; Smith, V. (2004). " Koh%20et%20al%202004%20extinction.pdf Species Coextinctions and the Biodiversity Crisis (http://www.unalmed.edu.co/~poboyca/documentos/documentos1/ Biología_Conservacion/03_2008/Polania/Set_03)". *Science* **305**: 1632-1634. doi: 10.1126/science.1101101 (http://dx.doi.org/10.1126/ science.1101101). Koh%20et%20al%202004%20extinction.pdf.

[231] Western, D. (1992). " The Biodiversity Crisis: A Challenge for Biology (http://www.jstor.org/pss/3545513)". *Oikos* **63** (1): 29-38. .

[232] Jackson JB (August 2008). " Colloquium paper: ecological extinction and evolution in the brave new ocean (http://www.pnas.org/cgi/ pmidlookup?view=long&pmid=18695220)". *Proc. Natl. Acad. Sci. U.S.A.* **105** (Suppl 1): 11458–65. doi: 10.1073/pnas.0802812105 (http:// dx.doi.org/10.1073/pnas.0802812105). PMID 18695220 (http://www.ncbi.nlm.nih.gov/pubmed/18695220). PMC 2556419 (http:// www.pubmedcentral.nih.gov/articlerender.fcgi?tool=pmcentrez&artid=2556419). .

[233] http://www.teebweb.org/

[234] http://ec.europa.eu/environment/nature/biodiversity/economics/

[235] Edwards, P. J.; Abivardi, C. (1998). " The value of biodiversity: Where ecology and economy blend (http://www.sciencedirect.com/ science?_ob=ArticleURL&_udi=B6V5X-3SX5K90-16&_user=10&_rdoc=1&_fmt=&_orig=search&_sort=d&_docanchor=&view=c& _searchStrId=1149767900&_rerunOrigin=scholar.google&_acct=C000050221&_version=0&_urlVersion=0&_userid=10& md5=45821b58d142650bebe9ec6939466dc5)". *Biological Conservation* **83** (2): 239-246. .

[236] Naidoo, R.; Malcolm, T.; Tomasek, A. (2009). " Economic benefits of standing forests in highland areas of Borneo: quantification and policy impacts (http://www.azoresbioportal.angra.uac.pt/files/publicacoes_Naiddo et al2009.pdf)". *Conservation Letters* **2**: 35-44. .

[237] Zhoua, X.; Al-Kaisib, M.; Helmers, M. J. (2009). " Cost effectiveness of conservation practices in controlling water erosion in Iowa (http:/ /www.sciencedirect.com/science?_ob=ArticleURL&_udi=B6TC6-4XHJX24-4&_user=10&_rdoc=1&_fmt=&_orig=search&_sort=d& _docanchor=&view=c&_searchStrId=1149771811&_rerunOrigin=scholar.google&_acct=C000050221&_version=1&_urlVersion=0& _userid=10&md5=5f9384acc1766d7349149976b750d60e)". *Soil and Tillage Research* **106** (1): 71-78. .

[238] Ferguson, K. (2006). " The True Value of Forests (http://www.jstor.org/pss/3868812)". *Frontiers in Ecology and the Environment* **4** (9): 456. .

[239] Anielski, M.; Wilson, S. (2005), *Counting Canada's Natural Capital: Assessing the Real value of Canada's Boreal Ecosystems* (http:// www.borealcanada.ca/pr_docs/Boreal_Wealth_Report_Nov_2005.pdf), Can. Bor. Ini., Pembina Institute, Ottawa,

[240] http://www.phschool.com/el_marketing.html

[241] http://www.pnas.org/content/96/21/11904.full

[242] http://www.nature.com/nature/journal/v426/n6968/full/426769a.html

[243] http://plato.

[244] http://scienceaid.co.uk/biology/ecology/index.html

[245] http://ekolojinet.com/journals.html

[246] http://ecologydictionary.org

Environmental geography

Environmental geography is the branch of geography that describes the spatial aspects of interactions between humans and the natural world. It requires an understanding of the dynamics of geology, meteorology, hydrology, biogeography, → ecology and geomorphology, as well as the ways in which human societies conceptualize the environment.

The links between cultural and physical geography were once more readily apparent than they are today. As human experience of the world is increasingly mediated by technology, the relationships have often become obscured.

Environmental geography represents a critically important set of analytical tools for assessing the impact of human presence on the environment by measuring the result of human activity on natural landforms and cycles.

Environmental geography is one of three, environmental, physical and human. Environmental concentrates on the relationship between human and the surrounding world.

See also

- Physical geography
- Meteorology
- Climate
- Human geography

Clearcutting

Clearcutting, or **clearfelling**, is a controversial forestry/logging practice in which most or all trees in a harvest area are cut down. Logging companies and forest-worker unions in some countries still support the practice for safety and economical reasons. Detractors see clearcutting as synonymous with deforestation, destroying natural habitats and contributing to global warming.

Clearcutting in Southern Finland

Types

Many variations of clearcutting exist, the most common professional practices are:

- *Standard (uniform) clearcut* – removal of every stem (whether commercially viable or not), so no canopy remains.
- *Patch clearcut* – removal of all the stems in a limited, predetermined area (patch).
- *Strip clearcut* – removal of all the stems in a row (strip), usually placed perpendicular to the prevailing winds in order to minimize the possibility of windthrow.
- *Clearcutting-with-reserves* – removal of the majority of standing stems save a few reserved for other purposes (for example as snags for wildlife habitat), (often confused with the seed tree method).
- Clearcutting contrasts with selective cutting such as high grading, in which only commercially valuable trees are harvested, leaving the others. This practice this can reduce the genetic viability of the forest over time, resulting in poorer or less vigorous offspring in the stand. Clearcutting also differs from a coppicing system, by allowing revegetation by seedlings.

Additionally, destructive forms of forest management are commonly referred to as 'clearcutting'.

- *Slash-and-burn* – the permanent conversion of tropical and subtropicals forests for agricultural purposes. This is most prevalent in tropical and subtropical forests in overpopulated regions in developing and least developed countries. Slash-and-burn entails the removal of all stems in a particular area. This is a form of deforestation, because the land is converted to other uses.
- Another controversial practice is the wholescale removal of irreplaceable ancient temperate and boreal forest.

Positive perspectives

Limited clearcutting can be practiced to encourage tree species that require high light intensity. Generally, a harvest area wider than double the height of the adjacent trees will no longer be subject to the moderating influence of the woodland on the microclimate.[1] The width of the harvest area can thus determine which species will come to dominate. Those with high tolerance to extremes in temperature, soil moisture, and resistance to browsing may be established, in particular secondary successional pioneer species.

Clearcutting is sometimes used by foresters as a method of mimicking disturbance and increasing primary successional species like poplar (aspen), willow and black cherry (North America). Clearcutting has also proved to be effective in creating animal habitat and browsing areas, which otherwise would not exist without natural stand replacing disturbances such as wildfires, large scale windthrow, or avalanches.

In temperate and boreal climates, clearcutting can have an effect on the depth of snow, which is usually greater in a clearcut area than in the forest, due to a lack of interception and evapotranspiration. This results in less soil frost, which in combination with higher levels of direct sunlight results in snowmelt occurring earlier in the spring.[2]

Negative impacts

Clearcutting can have major negative impacts. These have been cited as: poor quality re-growth, pest epidemics, increased wildfires, loss of biodiversity, loss of economic sustainability, loss of soil and increased environmental instability, and so on. Moreover the list of problems is growing. The response of governments around the world has been to marginalize community, environmental, social and academic concerns and defer to the logging industry.

In considerations of the almost universal practice of clear-cut forestry, governments have simply offered mundane facts in a manner similar to this. "Before the advent of modern forestry, high grading was the chief method of logging, with no regeneration for the areas cut, any re-growth converts to other uses or is left to regenerate naturally. In areas of the world where replanting is not undertaken, this continues to be the case. In the past and present, this kind of clearcutting without any replanting is practiced in forests where virtually every tree is valuable, as in an old-growth forest."

However, in practice old-growth wood often bears the scars af countless centuries, focusing immediate interest on optimally reproductive mid-growth trees, the forest powerhouses that in most market species are 200 to 500 yearts old. In recently undisturbed environments, old-growth forest really does shelter many optimal specimens that inspire clear-cut mentality as profit margins grow in the minds of reckless loggers. The deleterious effects of clear-cut on forestry are massive and multiple repairs have been attempted: three industry-wide and government supported attempts to improve clear-cut outlook include natural regrowth, careful selective seeding (monocropping) and burn-out. As with deforestation of primary forest, all repair attempts to date are poorly managed and sporadic, driven by profit, not unbiased forestry science. All attempts have failed. Another hopeful on the horizon is genetic manipulation.

Well organized and even costly repair attempts have turned into monumental failures. Where resource economies do not include purposeful seeding, the resulting clear-cut re-growth is always unhealthy and highly prone to massive fire damage. For example, in British Columbia, Canada where forestry is almost exclusively clear-cut practice, both the extent and the frequency of wild fire are rapidly increasing as a century of clear-cut, leveling 80% of the available forestry resource has resulted a huge biomass of flammable, unhealthy, and weedy fire fuel and pest food. In addition, in approximately one tenth of British Columbia forests, where biased industrial science dictated re-seeding to rule out "poor practice" then associated with unattended re-growth forests, complete destruction of tree-crop due to forest pests has resulted.

Clearcut re-seeding (that is improperly planned) has all of the same negative fire and disease effects of natural regrowth clearcuts. Another reseeding disadvantage, that is extensive where managed forestry companies exceed cuts and fill in slope margins with compulsory plantings, is mono-crop disease leading to soil failure. For instance, clearcutting on steep slopes always results high soil erosion rates. Tree species that can handle higher soil acidity (associated with soil erosion) and with roots suitable to retain and partially rebuild soil, such as pine trees in monocrop plantations. Unfortunately, extensive use of pine has wiped out millions of hectares of diseased and burned re-growth forests. The pine beetle larvae lacking the checks found in old growth and are now epidemic in North America. Pine beetles burrow into re-growth pine mono-cropping and kill the trees just when they enter their productive life-cycle point of rapid growth that should last 100 to 200 years. Dead clear-cut trees are accelerants in wildfires that are removing incresing amounts of the commercially useful vegetation in their path each year.

Burnout was not helpful, though it is the only option known to immediately protect property in the line of active fire. It was once thought that careful clear-cut burn-outs set to target just certain very congested patches and only certain undergrowth flammables would limit fire damage in expanding clear-cut zones. However, this did not limit wild fire incidence or extent across decades. In fact, it did create a huge cache of wild fire fuel, as forest cleaned by burnout become even more stressed by lack of biodiversity, soil erosion and disease.

Burn-outs are globally coupled with wildfire suppression. Suppression has disastrously increased re-growth fire fuel: by putting out wild fires, what burned small four decades ago becomes accelerant for much greater fuel mass during inevitable current and future fires.

Clearcutting in British Columbia

Clearcutting has created problems for British Columbia that many feel will persist, possibly for centuries. Much forestry science is certainly in agreement with that grim outlook, and government is currently guiding public relations and public education to manage that outlook. It is hoped that restricting human activity into and around clear-cut (80% of the Province of British Columbia) may lessen current outcomes for human fatality, fire incidence, and tree-crop loss.

There are no plans for regeneration of healthy forest, but forestry companies are turning to harvest natural regrowth every twenty years or so before fire and pests hit, providing a market in small wood objects like chop-sticks. No industrial attempt is made to mimic the healthy universal biodiversity of old-growth forest. When replants die out, the land is abandoned. To date, no government or science has costed and effectively assessed the complexity of restablishing biodiversity lost to clear-cut. Human beings simply complain about carnage or coddle concerns for a few isloated species. We do nothing meaningful to correct forest carnage. A few large corporations with a quota pointed at public placation have provided a few tiny plots of model old-growth in tourism venues. Governments claim to be protecting remote wildreness where a constant flow of cut timber is seen towed out to sea in huge carfully constructed ramps of tied logs.

Historically, human habitation has used clear-cut as a method to try and restrain a biosphere that would always recover within a few years and inundate human activity. Modern science and industrial methods have globally decimated the biodiversity that appears to sustain optimal health and survivability for life on earth. While we rush to make automobiles cleaner, we almost constantly fail to understand that this will have virtually no counter-effect equivalent to the massive global environmental damage resulting now from clear-cut activity. The single outlook for clear-cut is environmental devastation. We do not have the physical resource, the cultural controls or the science and technology to repair forests that we have turned into reservoirs of → extinction. Any image of scorched earth under mono-crop forest's black poles in British Columbia accurately illustrates just a very small parcel of the future environment for deforested earth's undirected human population. However, this reality seems only to fuel the technocratic economies destroying environment today. Clearcutting is the ultimate deadly global conundrum to which no words can deliver justice. Our future, the coming millenium so clearly invites images of failed human economy and culture in a global desert, treeless and barren.

See also

- Even-aged timber management
- List of tree species by shade tolerance – shade intolerant and some intermediate species are primarily regenerated with clearcuts
- Seed production and gene diversity

References

[1] Dr. J. Bowyer; K. Fernholz, A. Lindburg, Dr. J. Howe, Dr. S. Bratkovich (2009-05-28) (pdf). *The Power of Silviculture: Employing Thinning, Partial Cutting Systems and Other Intermediate Treatments to Increase Productivity, Forest Health and Public Support for Forestry* (http://dovetailinc.org/files/DovetailSilvics0509.pdf). Dovetail Partners Inc.. . Retrieved 2009-06-06.

[2] Ottosson Löfvenius, M.; Kluge, M., Lundmark, T.. (2003). "Snow and Soil Frost Depth in Two Types of Shelterwood and a Clear cut Area". *Scandinavian Journal of Forest Research* (Taylor & Francis) **18**: 54–63. ISSN 0282-7581 (http://worldcat.org/issn/0282-7581).

- Roy, Vincent, Ruelb, Jean-Claude and Plamondon, André P. (1999). 'Establishment, growth and survival of natural regeneration after clearcutting and drainage on forested wetlands' in *Forest Ecology and Management*, Volume 129, Issues 1-3, 17 April 2000, Pages 253-267 (http://www.sciencedirect.com/science?_ob=ArticleURL&_udi=B6T6X-3YRW05D-S&_user=10&_rdoc=1&_fmt=&_orig=search&_sort=d&_docanchor=&view=c&_searchStrId=1134686572&_rerunOrigin=google&_acct=C000050221&_version=1&_urlVersion=0&_userid=10&md5=a20512852a24923befd2b797f8db6133)

External links

- Belt, Kevin and Campbell, Robert, *The Clearcutting Controversy - Myths and Facts*, West Virginia University, accessed 14 December 2009 (http://www.wvu.edu/~agexten/forestry/clrcut.htm)
- CBC Digital Archives - Clearcutting and Logging: The War of the Woods (http://archives.cbc.ca/IDD-1-75-679/science_technology/clearcutting/)
- Clearcutting in Maine (http://www.umaine.edu/mcsc/MPR/Vol5No2/hagen.htm)
- Clearcutting (http://www.ritchiewiki.com/wiki/index.php/Clearcutting) on the Ritchiewiki
- Congressional Research Service (CRS) Reports regarding Clearcutting, accessed 14 December 2009 (http://digital.library.unt.edu/govdocs/crs/search/?q=clearcutting&t=fulltext&nlow=&nhi=)
- Forest Policy Research page: California citizens to stop Sierra Pacifics plan to clearcut one million acres of Sierra forest, accessed 14 December 2009 (http://forestpolicyresearch.org/2009/03/04/california-citizens-to-stop-sierra-pacifics-plan-to-clearcut-one-million-acres-of-sierra-forest/)

Internal combustion engine

The **internal combustion engine** is an engine in which the combustion of a fuel (generally, fossil fuel) occurs with an oxidizer (usually air) in a combustion chamber. In an internal combustion engine the expansion of the high temperature and pressure gases, which are produced by the combustion, directly applies force to a movable component of the engine, such as the pistons or turbine blades and by moving it over a distance, generate useful mechanical energy.[1] [2] [3] [4]

An automobile engine partly opened and colored
to show components

The term *internal combustion engine* usually refers to an engine in which combustion is intermittent, such as the more familiar four-stroke and two-stroke piston engines, along with variants, such as the Wankel rotary engine. A second class of internal combustion engines use continuous combustion: gas turbines, jet engines and most rocket engines, each of which are internal combustion engines on the same principle as previously described.[1] [2] [3] [4]

The internal combustion engine (or ICE) is quite different from external combustion engines, such as steam or Stirling engines, in which the energy is delivered to a working fluid not consisting of, mixed with or contaminated by combustion products. Working fluids can be air, hot water, pressurised water or even liquid sodium, heated in some kind of boiler by fossil fuel, wood-burning, nuclear, solar etc.

A large number of different designs for ICEs have been developed and built, with a variety of different strengths and weaknesses. Powered by an energy-dense fuel (which is very frequently petrol, a liquid derived from fossil fuels) the ICE delivers an excellent power-to-weight ratio with few safety or other disadvantages. While there have been and still are many stationary applications, the real strength of internal combustion engines is in mobile applications and they dominate as a power supply for cars, aircraft, and boats, from the smallest to the biggest. Only for hand-held power tools do they share part of the market with battery powered devices.

Applications

Internal combustion engines are most commonly used for mobile propulsion in vehicles and portable machinery. In mobile equipment, internal combustion is advantageous since it can provide high power-to-weight ratios together with excellent fuel energy density. Generally using fossil fuel (mainly petroleum), these engines have appeared in transport in almost all vehicles (automobiles, trucks, motorcycles, boats, and in a wide variety of aircraft and locomotives).

Internal combustion engines appear in the form of gas turbines as well where a very high power is required, such as in jet aircraft, helicopters, and large ships. They are also frequently used for electric generators and by industry.

A 1906 gasoline engine

Classification

At one time the word, "Engine" (from Latin, via Old French, *ingenium*, "ability") meant any piece of machinery—a sense that persists in expressions such as *siege engine*. A "motor" (from Latin *motor*, "mover") is any machine that produces mechanical power. Traditionally, electric motors are not referred to as "Engines"; however, combustion engines are often referred to as "motors." (An *electric engine* refers to a locomotive operated by electricity.)

Engines can be classified in many different ways: By the engine cycle used, the layout of the engine, source of energy, the use of the engine, or by the cooling system employed.

Principles of operation

Reciprocating:

- Two-stroke cycle
- Four-stroke cycle
- Six-stroke engine
- Diesel engine
- Atkinson cycle

Rotary:

- Wankel engine

Continuous combustion:

Brayton cycle:

- Gas turbine
- Jet engine (including turbojet, turbofan, ramjet, Rocket etc.)

Engine configurations

Internal combustion engines can be classified by their configuration.

Four stroke configuration

Operation

Basic process

As their name implies, operation of a four stroke internal combustion engines have 4 basic steps that repeat with every two revolutions of the engine:

1. **Intake**
   - Combustible mixtures are emplaced in the combustion chamber
2. **Compression**
   - The mixtures are placed under pressure
3. **Power**
   - The mixture is burnt, almost invariably a *deflagration*, although a few systems involve *detonation*. The hot mixture is expanded, pressing on and moving parts of the engine and performing useful work.
4. **Exhaust**
   - The cooled combustion products are exhausted into the atmosphere

Many engines overlap these steps in time; jet engines do all steps simultaneously at different parts of the engines.

Combustion

All **internal combustion engines** depend on the exothermic chemical process of combustion: the reaction of a fuel, typically with oxygen from the air (though it is possible to inject nitrous oxide in order to do more of the same thing and gain a power boost). The combustion process typically results in the production of a great quantity of heat, as well as the production of steam and carbon dioxide and other chemicals at very high temperature; the temperature reached is determined by the chemical make up of the fuel and oxidisers (see stoichiometry).

Four-stroke cycle (or Otto cycle)
1. Intake
2. Compression
3. Power
4. Exhaust

The most common modern fuels are made up of hydrocarbons and are derived mostly from fossil fuels (petroleum). Fossil fuels include diesel fuel, gasoline and petroleum gas, and the rarer use of propane. Except for the fuel delivery components, most internal combustion engines that are designed for gasoline use can run on natural gas or liquefied petroleum gases without major modifications. Large diesels can run with air mixed with gases and a pilot diesel fuel ignition injection. Liquid and gaseous biofuels, such as ethanol and biodiesel (a form of diesel fuel that is produced from crops that yield triglycerides such as soybean oil), can also be used. Some engines with appropriate modifications can also run on hydrogen gas.

Internal combustion engines require ignition of the mixture, either by spark ignition (SI) or compression ignition (CI). Before the invention of reliable electrical methods, hot tube and flame methods were used.

Gasoline Ignition Process

Gasoline engine ignition systems generally rely on a combination of a lead-acid battery and an induction coil to provide a high-voltage electrical spark to ignite the air-fuel mix in the engine's cylinders. This battery is recharged during operation using an electricity-generating device such as an alternator or generator driven by the engine. Gasoline engines take in a mixture of air and gasoline and compress it to not more than 12.8 bar (1.28 MPa), then use a spark plug to ignite the mixture when it is compressed by the piston head in each cylinder.

Diesel Ignition Process

Diesel engines and HCCI (Homogeneous charge compression ignition) engines, rely solely on heat and pressure created by the engine in its compression process for ignition. The compression level that occurs is usually twice or more than a gasoline engine. Diesel engines will take in air only, and shortly before peak compression, a small quantity of diesel fuel is sprayed into the cylinder via a fuel injector that allows the fuel to instantly ignite. HCCI type engines will take in both air and fuel but continue to rely on an unaided auto-combustion process, due to higher pressures and heat. This is also why diesel and HCCI engines are more susceptible to cold-starting issues, although they will run just as well in cold weather once started. Light duty diesel engines with indirect injection in automobiles and light trucks employ glowplugs that pre-heat the combustion chamber just before starting to reduce no-start conditions in cold weather. Most diesels also have a battery and charging system; nevertheless, this system is secondary and is added by manufacturers as a luxury for the ease of starting, turning fuel on and off (which can also be done via a switch or mechanical apparatus), and for running auxiliary electrical components and accessories. Most new engines rely on electrical and electronic control system that also control the combustion process to increase efficiency and reduce emissions.

Two stroke configuration

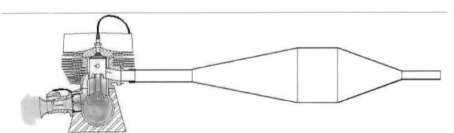

Animated two stroke engine in operation

Engines based on the two-stroke cycle use two strokes (one up, one down) for every power stroke. Since there are no dedicated intake or exhaust strokes, alternative methods must be used to scavenge the cylinders. The most common method in spark-ignition two-strokes is to use the downward motion of the piston to pressurize fresh charge in the crankcase, which is then blown through the cylinder through ports in the cylinder walls.

Spark-ignition two-strokes are small and light for their power output and mechanically very simple; however, they are also generally less efficient and more polluting than their four-stroke counterparts. In terms of power per cubic centimetre, a single-cylinder small motor application like a two-stroke engine produces much more power than an equivalent four-stroke engine due to the enormous advantage of having one power stroke for every 360 degrees of crankshaft rotation (compared to 720 degrees in a 4 stroke motor).

Small displacement, crankcase-scavenged two-stroke engines have been less fuel-efficient than other types of engines when the fuel is mixed with the air prior to scavenging allowing some of it to escape out of the exhaust port. Modern designs (Sarich and Paggio) use air-assisted fuel injection which avoids this loss, and are more efficient than comparably sized four-stroke engines. Fuel injection is essential for a modern two-stroke engine in order to meet ever more stringent emission standards.

Research continues into improving many aspects of two-stroke motors including direct fuel injection, amongst other things. The initial results have produced motors that are much cleaner burning than their traditional counterparts. Two-stroke engines are widely used in snowmobiles, lawnmowers, string trimmers, chain saws, jet skis, mopeds, outboard motors, and many motorcycles. Two-stroke engines have the advantage of an increased specific power ratio (i.e. *power to volume ratio*), typically around 1.5 times that of a typical four-stroke engine.

The largest compression-ignition engines are two-strokes and are used in some locomotives and large ships. These particular engines use forced induction to scavenge the cylinders; an example of this type of motor is the Wartsila-Sulzer turbocharged two-stroke diesel as used in large container ships. It is the most efficient and powerful internal combustion engine in the world with over 50% thermal efficiency. For comparison, the most efficient small four-stroke motors are around 43% thermal efficiency (SAE 900648); size is an advantage for efficiency due to the increase in the ratio of volume to surface area.

Common cylinder configurations include the straight or inline configuration, the more compact V configuration, and the wider but smoother flat or boxer configuration. Aircraft engines can also adopt a radial configuration which allows more effective cooling. More unusual configurations such as the H, U, X, and W have also been used.

Multiple crankshaft configurations do not necessarily need a cylinder head at all because they can instead have a piston at each end of the cylinder called an opposed piston design. Because here gas in- and outlets are positioned at opposed ends of the cylinder, one can achieve uniflow scavenging, which is, like in the four stroke engine, efficient over a wide range of revolution numbers. Also the thermal efficiency is improved because of lack of cylinder heads. This design was used in the Junkers Jumo 205 diesel aircraft engine, using at either end of a single bank of cylinders with two crankshafts, and most remarkably in the Napier Deltic diesel engines. These used three crankshafts to serve three banks of double-ended cylinders arranged in an equilateral triangle with the crankshafts at the corners. It was also used in single-bank locomotive engines, and continues to be used for marine engines, both for propulsion and for auxiliary generators. The Gnome Rotary engine, used in several early aircraft, had a stationary crankshaft and a bank of radially arranged cylinders rotating around it.

Wankel

The Wankel engine (rotary engine) does not have piston strokes. It operates with the same separation of phases as the four-stroke engine with the phases taking place in separate locations in the engine. In thermodynamic terms it follows the Otto engine cycle, so may be thought of as a "four-phase" engine. While it is true that three power strokes typically occur per rotor revolution due to the 3/1 revolution ratio of the rotor to the eccentric shaft, only one power stroke per shaft revolution actually occurs; this engine provides three power 'strokes' per revolution per rotor giving it a greater power-to-weight ratio than piston engines. This type of engine is most notably used in the current Mazda RX-8, the earlier RX-7, and other models.

Gas turbines

A gas turbine is a rotary machine similar in principle to a steam turbine and it consists of three main components: a compressor, a combustion chamber, and a turbine. The air after being compressed in the compressor is heated by burning fuel in it. About two-thirds of the

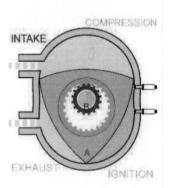

The Wankel cycle. The shaft turns three times for each rotation of the rotor around the lobe and once for each orbital revolution around the eccentric shaft.

heated air combined with the products of combustion is expanded in a turbine resulting in work output which is used to drive the compressor. The rest (about one-third) is available as useful work output.

Jet engine

Jet engines take a large volume of hot gas from a combustion process (typically a gas turbine, but rocket forms of jet propulsion often use solid or liquid propellants, and ramjet forms also lack the gas turbine) and feed it through a nozzle which accelerates the jet to high speed. As the jet accelerates through the nozzle, this creates thrust and in turn does useful work.

Engine cycle

Two-stroke

This system manages to pack one power stroke into every two strokes of the piston (up-down). This is achieved by exhausting and re-charging the cylinder simultaneously.

The steps involved here are:

1. Intake and exhaust occur at bottom dead center. Some form of pressure is needed, either crankcase compression or super-charging.
2. Compression stroke: Fuel-air mix compressed and ignited. In case of Diesel: Air compressed, fuel injected and self ignited
3. Power stroke: piston is pushed downwards by the hot exhaust gases.

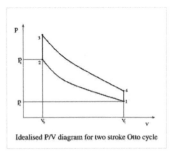

Idealised P/V diagram for two stroke Otto cycle

Four-stroke

Engines based on the four-stroke ("Otto cycle") have one power stroke for every four strokes (up-down-up-down) and employ spark plug ignition. Combustion occurs rapidly, and during combustion the volume varies little ("constant volume").[5] They are used in cars, larger boats, some motorcycles, and many light aircraft. They are generally quieter, more efficient, and larger than their two-stroke counterparts.

The steps involved here are:

1. Intake stroke: Air and vaporized fuel are drawn in.
2. Compression stroke: Fuel vapor and air are compressed and ignited.
3. Combustion stroke: Fuel combusts and piston is pushed downwards.
4. Exhaust stroke: Exhaust is driven out. During the 1st, 2nd, and 4th

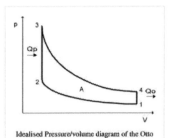

Idealised Pressure/volume diagram of the Otto cycle showing combustion heat input Qp and waste exhaust output Qo, the power stroke is the top curved line, the bottom is the compression stroke

stroke the piston is relying on power and the momentum generated by the other pistons. In that case, a four cylinder engine would be less powerful than a six or eight cylinder engine.

There are a number of variations of these cycles, most notably the Atkinson and Miller cycles. The diesel cycle is somewhat different.

Diesel cycle

Most truck and automotive diesel engines use a cycle reminiscent of a four-stroke cycle, but with a compression heating ignition system, rather than needing a separate ignition system. This variation is called the diesel cycle. In the diesel cycle, diesel fuel is injected directly into the cylinder so that combustion occurs at constant pressure, as the piston moves, rather than with the four stroke with the piston essentially stationary.

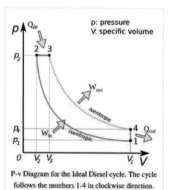

P-v Diagram for the Ideal Diesel cycle. The cycle follows the numbers 1-4 in clockwise direction.

Six-stroke

The six-stroke engine captures the wasted heat from the four-stroke Otto cycle and creates steam, which simultaneously cools the engine while providing a free power stroke. This removes the need for a cooling system making the engine lighter while giving 40% increased efficiency over the Otto Cycle.

Brayton cycle

A gas turbine is a rotary machine somewhat similar in principle to a steam turbine and it consists of three main components: a compressor, a combustion chamber, and a turbine. The air after being compressed in the compressor is heated by burning fuel in it, this heats and expands the air, and this extra energy is tapped by the turbine which in turn powers the compressor closing the cycle and powering the shaft.

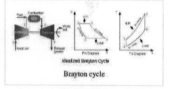

Brayton cycle

Gas turbine cycle engines employ a continuous combustion system where compression, combustion, and expansion occur simultaneously at different places in the engine—giving continuous power. Notably the combustion takes place at constant pressure, rather than with the Otto cycle, constant volume.

Disused methods

In some old noncompressing internal combustion engines: in the first part of the piston downstroke, a fuel-air mixture was sucked or blown in, and in the rest of the piston downstroke, the inlet valve closed and the fuel-air mixture fired. In the piston upstroke, the exhaust valve was open. This was an attempt at imitating the way a piston steam engine works, and since the explosive mixture was not compressed, the heat and pressure generated by combustion was much less causing lower overall efficiency.

Fuels and oxidizers

Engines are often classified by the fuel (or propellant) used.

Fuels

Nowadays, fuels used include:

- Petroleum:
  - Petroleum spirit (North American term: gasoline, British term: petrol)
  - Petroleum diesel.
  - Autogas (liquified petroleum gas).
  - Compressed natural gas.
  - Jet fuel (aviation fuel)
  - Residual fuel
- Coal:
  - Most methanol is made from coal.
  - Gasoline can be made from carbon (coal) using the Fischer-Tropsch process
  - Diesel fuel can be made from carbon using the Fischer-Tropsch process
- Biofuels and vegoils:
  - Peanut oil and other vegoils.
  - Biofuels:
    - Biobutanol (replaces gasoline).
    - Biodiesel (replaces petrodiesel).
    - Bioethanol and Biomethanol (wood alcohol) and other biofuels (see Flexible-fuel vehicle).
    - Biogas
- Hydrogen (mainly spacecraft rocket engines)

Even fluidized metal powders and explosives have seen some use. Engines that use gases for fuel are called gas engines and those that use liquid hydrocarbons are called oil engines, however gasoline engines are also often colloquially referred to as, "gas engines" ("petrol engines" in the UK).

The main limitations on fuels are that it must be easily transportable through the fuel system to the combustion chamber, and that the fuel releases sufficient energy in the form of heat upon combustion to make practical use of the engine.

Diesel engines are generally heavier, noisier, and more powerful at lower speeds than gasoline engines. They are also more fuel-efficient in most circumstances and are used in heavy road vehicles, some automobiles (increasingly so for their increased fuel efficiency over gasoline engines), ships, railway locomotives, and light aircraft. Gasoline engines are used in most other road vehicles including most cars, motorcycles, and mopeds. Note that in Europe, sophisticated diesel-engined cars have taken over about 40% of the market since the 1990s. There are also engines that run on hydrogen, methanol, ethanol, liquefied petroleum gas (LPG), and biodiesel. Paraffin and tractor vaporizing oil (TVO) engines are no longer seen.

Hydrogen

At present, hydrogen is mostly used as fuel for rocket engines. In the future, hydrogen might replace more conventional fuels in traditional internal combustion engines. If hydrogen fuel cell technology becomes widespread, then the use of internal combustion engines may be phased out.

Although there are multiple ways of producing free hydrogen, those methods require converting combustible molecules into hydrogen or consuming electric energy. Unless that electricity is produced from a renewable source—and is not required for other purposes— hydrogen does not solve any energy crisis. In many situations the disadvantage of hydrogen, relative to carbon fuels, is its storage. Liquid hydrogen has extremely low density (14 times lower than water) and requires extensive insulation—whilst gaseous hydrogen requires heavy tankage. Even when liquefied, hydrogen has a higher specific energy but the volumetric energetic storage is still roughly five times lower than petrol. However the energy density of hydrogen is considerably higher than that of electric batteries, making it a serious contender as an energy carrier to replace fossil fuels. The 'Hydrogen on Demand' process (see direct borohydride fuel cell) creates hydrogen as it is needed, but has other issues such as the high price of the sodium borohydride which is the raw material.

Oxidizers

Since air is plentiful at the surface of the earth, the oxidizer is typically atmospheric oxygen which has the advantage of not being stored within the vehicle, increasing the power-to-weight and power to volume ratios. There are other materials that are used for special purposes, often to increase power output or to allow operation under water or in space.

- Compressed air has been commonly used in torpedoes.
- Compressed oxygen, as well as some compressed air, was used in the Japanese Type 93 torpedo. Some submarines are designed to carry pure oxygen. Rockets very often use liquid oxygen.
- Nitromethane is added to some racing and model fuels to increase power and control combustion.
- Nitrous oxide has been used—with extra gasoline—in tactical aircraft and in specially equipped cars to allow short bursts of added power from engines that otherwise run on gasoline and air. It is also used in the Burt Rutan rocket spacecraft.
- Hydrogen peroxide power was under development for German World War II submarines and may have been used in some non-nuclear submarines and was used on some rocket engines (notably Black Arrow and Me-163 rocket plane)
- Other chemicals such as chlorine or fluorine have been used experimentally, but have not been found to be practical.

One-cylinder gasoline engine (ca. 1910).

Engine capacity

For piston engines, an engine's capacity is the engine displacement, in other words the volume swept by all the pistons of an engine in a single movement. It is generally measured in litres (L) or cubic inches (c.i.d. *or* cu in *or* in³) for larger engines, and cubic centimetres (abbreviated cc) for smaller engines. Engines with greater capacities are usually more powerful and provide greater torque at lower rpm, but also consume more fuel. Apart from designing an engine with more cylinders, there are two ways to increase an engines' capacity. The first is to lengthen the stroke: the second is to increase the pistons' diameter *(See also: Stroke ratio)*. In either case, it may be necessary to make further adjustments to the fuel intake of the engine to ensure optimum performance.

Common components

Combustion chambers

Internal combustion engines can contain any number of combustion chambers (cylinders), with numbers between one and twelve being common, though as many as 36 (Lycoming R-7755) have been used. Having more cylinders in an engine yields two potential benefits: first, the engine can have a larger displacement with smaller individual reciprocating masses, that is, the mass of each piston can be less thus making a smoother-running engine since the engine tends to vibrate as a result of the pistons moving up and down. Doubling the number of the same size cylinders will double the torque and power. The downside to having more pistons is that the engine will tend to weigh more and generate more internal friction as the greater number of pistons rub against the inside of their cylinders. This tends to decrease fuel efficiency and robs the engine of some of its power. For high-performance gasoline engines using current materials and technology—such as the engines found in modern automobiles, there seems to be a break-point around 10 or 12 cylinders after which the addition of cylinders becomes an overall detriment to performance and efficiency. Although, exceptions such as the W16 engine from Volkswagen exist.

- Most car engines have four to eight cylinders with some high performance cars having ten, twelve—or even sixteen, and some very small cars and trucks having two or three. In previous years, some quite large cars such as the DKW and Saab 92, had two-cylinder or two-stroke engines.
- Radial aircraft engines (now obsolete) had from three to 28 cylinders; an example is the Pratt & Whitney R-4360. A row contains an odd number of cylinders so an even number indicates a two- or four-row engine. The largest of these was the Lycoming R-7755 with 36 cylinders (four rows of nine cylinders), but it did not enter production.
- Motorcycles commonly have from one to four cylinders, with a few high performance models having six; although, some 'novelties' exist with 8, 10, or 12.
- Snowmobiles Usually have one to four cylinders and can be both 2 stroke or 4 stroke, normally in the in-line configuration however there are again some novelties that exist with V-4 Engines
- Small portable appliances such as chainsaws, generators, and domestic lawn mowers most commonly have one cylinder, but two-cylinder chainsaws exist.
- Large reversible two cycle marine diesels have a minimum of three to over ten cylinders. Freight diesel locomotives usually have around 12 to 20 cylinders due to space limitations as larger cylinders take more space (volume) per kwh, due to the limit on average piston speed of less than 30 ft/sec on engines lasting more than 40000 hours under full power.

Ignition system

The ignition system of an internal combustion engines depends on the type of engine and the fuel used. Petrol engines are typically ignited by a precisely timed spark, and diesel engines by compression heating. Historically, outside flame and hot-tube systems were used, see hot bulb engine.

Spark

The mixture is ignited by an electrical spark from a spark plug—the timing of which is very precisely controlled. Almost all gasoline engines are of this type. Diesel engines timing is precisely controlled by the pressure pump and injector.

Compression

Ignition occurs as the temperature of the fuel/air mixture is taken over its autoignition temperature, due to heat generated by the compression of the air during the compression stroke. The vast majority of compression ignition engines are diesels in which the fuel is mixed with the air after the air has reached ignition temperature. In this case, the timing comes from the fuel injection system. Very small model engines for which simplicity and light weight is more important than fuel costs use easily ignited fuels (a mixture of kerosene, ether, and lubricant) and adjustable compression to control ignition timing for starting and running.

Ignition timing

For reciprocating engines, the point in the cycle at which the fuel-oxidizer mixture is ignited has a direct effect on the efficiency and output of the ICE. The thermodynamics of the idealized Carnot heat engine tells us that an ICE is most efficient if most of the burning takes place at a high temperature, resulting from compression—near top dead center. The speed of the flame front is directly affected by the compression ratio, fuel mixture temperature, and Octane rating or cetane number of the fuel. Leaner mixtures and lower mixture pressures burn more slowly requiring more advanced ignition timing. It is important to have combustion spread by a thermal flame front (deflagration), not by a shock wave. Combustion propagation by a shock wave is called detonation and, in engines, is also known as pinging or Engine knocking.

So at least in gasoline-burning engines, ignition timing is largely a compromise between an earlier "advanced" spark—which gives greater efficiency with high octane fuel—and a later "retarded" spark that avoids detonation with the fuel used. For this reason, high-performance diesel automobile proponents such as, Gale Banks, believe that

> There's only so far you can go with an air-throttled engine on 91-octane gasoline. In other words, it is the fuel, gasoline, that has become the limiting factor. ... While turbocharging has been applied to both gasoline and diesel engines, only limited boost can be added to a gasoline engine before the fuel octane level again becomes a problem. With a diesel, boost pressure is essentially unlimited. It is literally possible to run as much boost as the engine will physically stand before breaking apart. Consequently, engine designers have come to realize that diesels are capable of substantially more power and torque than any comparably sized gasoline engine.[6]

Fuel systems

Fuels burn faster and more efficiently when they present a large surface area to the oxygen in air. Liquid fuels must be atomized to create a fuel-air mixture, traditionally this was done with a carburetor in petrol engines and with fuel injection in diesel engines. Most modern petrol engines now use fuel injection too - though the technology is quite different. While diesel must be injected at an exact point in that engine cycle, no such precision is needed in a petrol engine. However, the lack of lubricity in petrol means that the injectors themselves must be more sophisticated.

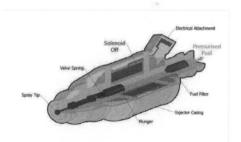

Animated cut through diagram of a typical fuel injector, a device used to deliver fuel to the internal combustion engine.

Carburetor

Simpler reciprocating engines continue to use a carburetor to supply fuel into the cylinder. Although carburetor technology in automobiles reached a very high degree of sophistication and precision, from the mid-1980s it lost out on cost and flexibility to fuel injection. Simple forms of carburetor remain in widespread use in small engines such as lawn mowers and more sophisticated forms are still used in small motorcycles.

Fuel injection

Larger gasoline engines used in automobiles have mostly moved to fuel injection systems (see Gasoline Direct Injection). Diesel engines have always used fuel injection system because the timing of the injection initiates and controls the combustion.

Autogas (LPG) engines use either fuel injection systems or open- or closed-loop carburetors.

Fuel pump

Most internal combustion engines now require a fuel pump. Diesel engines use an all-mechanical precision pump system that delivers a timed injection direct into the combustion chamber, hence requiring a high delivery pressure to overcome the pressure of the combustion chamber. Petrol fuel injection delivers into the inlet tract at atmospheric pressure (or below) and timing is not involved, these pumps are normally driven electrically. Gas turbine and rocket engines use electrical systems.

Other

Other internal combustion engines like jet engines and rocket engines employ various methods of fuel delivery including impinging jets, gas/liquid shear, preburners and others.

Oxidiser-Air inlet system

Some engines such as solid rockets have oxidisers already within the combustion chamber but in most cases for combustion to occur, a continuous supply of oxidiser must be supplied to the combustion chamber.

Natural aspirated engines

When air is used with piston engines it can simply suck it in as the piston increases the volume of the chamber. However, this gives a maximum of 1 atmosphere of pressure difference across the inlet valves, and at high engine speeds the resulting airflow can limit potential power output.

Superchargers and turbochargers

A supercharger is a "forced induction" system which uses a compressor powered by the shaft of the engine which forces air through the valves of the engine to achieve higher flow. When these systems are employed the maximum absolute pressure at the inlet valve is typically around 2 times atmospheric pressure or more.

Turbochargers are another type of forced induction system which has its compressor powered by a gas turbine running off the exhaust gases from the engine.

Turbochargers and superchargers are particularly useful at high altitudes and they are frequently used in aircraft engines.

Duct jet engines use the same basic system, but eschew the piston engine, and replace it with a burner instead.

A cutaway of a turbocharger

Liquids

In liquid rocket engines, the oxidiser comes in the form of a liquid and needs to be delivered at high pressure (typically 10-230 bar or 1–23 MPa) to the combustion chamber. This is normally achieved by the use of a centrifugal pump powered by a gas turbine - a configuration known as a *turbopump*, but it can also be pressure fed.

Parts

For a four-stroke engine, key parts of the engine include the crankshaft (purple), connecting rod (orange), one or more camshafts (red and blue), and valves. For a two-stroke engine, there may simply be an exhaust outlet and fuel inlet instead of a valve system. In both types of engines there are one or more cylinders (grey and green), and for each cylinder there is a spark plug (darker-grey, gasoline engines only), a piston (yellow), and a crankpin (purple). A single sweep of the cylinder by the piston in an upward or downward motion is known as a stroke. The downward stroke that occurs directly after the air-fuel mix passes from the carburetor or fuel injector to the cylinder (where it is ignited) is also known as a power stroke.

A Wankel engine has a triangular rotor that orbits in an epitrochoidal (figure 8 shape) chamber around an eccentric shaft. The four phases of operation (intake, compression, power, and exhaust) take place in what is effectively a moving, variable-volume chamber.

An illustration of several key components in a typical four-stroke engine.

Valves

All four-stroke internal combustion engines employ valves to control the admittance of fuel and air into the combustion chamber. Two-stroke engines use ports in the cylinder bore, covered and uncovered by the piston, though there have been variations such as exhaust valves.

Piston engine valves

In piston engines, the valves are grouped into 'inlet valves' which admit the entrance of fuel and air and 'outlet valves' which allow the exhaust gases to escape. Each valve opens once per cycle and the ones that are subject to extreme accelerations are held closed by springs that are typically opened by rods running on a camshaft rotating with the engines' crankshaft.

Control valves

Continuous combustion engines—as well as piston engines—usually have valves that open and close to admit the fuel and/or air at the startup and shutdown. Some valves feather to adjust the flow to control power or engine speed as well.

Exhaust systems

Internal combustion engines have to manage the exhaust of the cooled combustion gas from the engine. The exhaust system frequently contains devices to control pollution, both chemical and noise pollution. In addition, for cyclic combustion engines the exhaust system is frequently tuned to improve emptying of the combustion chamber.

For jet propulsion internal combustion engines, the 'exhaust system' takes the form of a high velocity nozzle, which generates thrust for the engine and forms a colimated jet of gas that gives the engine its name.

Cooling systems

Combustion generates a great deal of heat, and some of this transfers to the walls of the engine. Failure will occur if the body of the engine is allowed to reach too high a temperature; either the engine will physically fail, or any lubricants used will degrade to the point that they no longer protect the engine.

Cooling systems usually employ air (air cooled) or liquid (usually water) cooling while some very hot engines using radiative cooling (especially some Rocket engines). Some high altitude rocket engines use ablative cooling where the walls gradually erode in a controlled fashion. Rockets in particular can use regenerative cooling which uses the fuel to cool the solid parts of the engine.

Piston

A **piston** is a component of reciprocating engines. It is located in a cylinder and is made gas-tight by piston rings. Its purpose is to transfer force from expanding gas in the cylinder to the crankshaft via a piston rod and/or connecting rod. In two-stroke engines the piston also acts as a valve by covering and uncovering ports in the cylinder wall.

Propelling nozzle

For jet engine forms of internal combustion engines a propelling nozzle is present. This takes the high temperature, high pressure exhaust and expands and cools it. The exhaust leaves the nozzle going at much higher speed and provides thrust, as well as constricting the flow from the engine and raising the pressure in the rest of the engine, giving greater thrust for the exhaust mass that exits.

Crankshaft

Most reciprocating internal combustion engines end up turning a shaft. This means that the linear motion of a piston must be converted into rotation. This is typically achieved by a crankshaft.

Flywheels

The flywheel is a disk or wheel attached to the crank, forming an inertial mass that stores rotational energy. In engines with only a single cylinder the flywheel is essential to carry energy over from the power stroke into a subsequent compression stroke. Flywheels are present in most reciprocating engines to smooth out the power delivery over each rotation of the crank and in most automotive engines also mount a gear ring for a starter. The rotational inertia of the flywheel also allows a much slower minimum unloaded speed and also improves the smoothness at idle. The flywheel may also perform a part of the balancing of the system and so by itself be out of balance, although most engines will use a neutral balance for the flywheel, enabling it to be balanced in a separate operation. The flywheel is also used as a mounting for the clutch or a torque converter in most automotive applications.

Starter systems

All internal combustion engines require some form of system to get them into operation. Most piston engines use a starter motor powered by the same battery as runs the rest of the electric systems. Large jet engines and gas turbines are started with a compressed air motor that is geared to one of the engine's driveshafts. Compressed air can be supplied from another engine, a unit on the ground or by the aircraft's APU. Small internal combustion engines are often started by pull cords. Motorcycles of all sizes were traditionally kick-started, though all but the smallest are now electric-start. Large stationary and marine engines may be started by the timed injection of compressed air into the cylinders - or occasionally with cartridges. Jump starting refers to assistance from another battery (typically when the fitted battery is discharged), while bump starting refers to an alternative method of starting by the application of some external force, e.g. rolling down a hill.

Lubrication Systems

Internal combustions engines require lubrication in operation that moving parts slide smoothly over each other. Insufficient lubrication subjects the parts of the engine to metal-to-metal contact, friction, heat build-up, rapid wear often culminating in parts becoming friction welded together eg pistons in their cylinders. Big end bearings seizing up will sometimes lead to a connecting rod breaking and poking out through the crankcase.

Several different types of lubrication systems are used. Simple two-stroke engines are lubricated by oil mixed into the fuel or injected into the induction stream as a spray. Early slow-speed stationary and marine engines were lubricated by gravity from small chambers similar to those used on steam engines at the time—with an engine tender refilling these as needed. As engines were adapted for automotive and aircraft use, the need for a high power-to-weight ratio led to increased speeds, higher temperatures, and greater pressure on bearings which in turn required pressure-lubrication for crank bearings and connecting-rod journals. This was provided either by a direct lubrication from a pump, or indirectly by a jet of oil directed at pickup cups on the connecting rod ends which had the advantage of providing higher pressures as the engine speed increased.

Control systems

Most engines require one or more systems to start and shutdown the engine and to control parameters such as the power, speed, torque, pollution, combustion temperature, efficiency and to stabilise the engine from modes of operation that may induce self-damage such as pre-ignition. Such systems may be referred to as engine control units.

Many control systems today are digital, and are frequently termed FADEC (Full Authority Digital Electronic Control) systems.

Diagnostic systems

Engine On Board Diagnostics (also known as OBD) is a computerized system that allows for electronic diagnosis of a vehicles' powerplant. The first generation, known as *OBD1*, was introduced 10 years after the U.S. Congress passed the Clean Air Act in 1970 as a way to monitor a vehicles' fuel injection system. *OBD2*, the second generation of computerized on-board diagnostics, was codified and recommended by the California Air Resource Board in 1994 and became mandatory equipment aboard all vehicles sold in the United States as of 1996.

Measures of engine performance

Engine types vary greatly in a number of different ways:

- energy efficiency
- fuel/propellant consumption (brake specific fuel consumption for shaft engines, thrust specific fuel consumption for jet engines)
- power to weight ratio
- thrust to weight ratio
- Torque curves (for shaft engines) thrust lapse (jet engines)
- Compression ratio for piston engines, Overall pressure ratio for jet engines and gas turbines

Energy efficiency

Once ignited and burnt, the combustion products—hot gases—have more available thermal energy than the original compressed fuel-air mixture (which had higher chemical energy). The available energy is manifested as high temperature and pressure that can be translated into work by the engine. In a reciprocating engine, the high-pressure gases inside the cylinders drive the engine's pistons.

Once the available energy has been removed, the remaining hot gases are vented (often by opening a valve or exposing the exhaust outlet) and this allows the piston to return to its previous position (top dead center, or TDC). The piston can then proceed to the next phase of its cycle, which varies between engines. Any heat that isn't translated into work is normally considered a waste product and is removed from the engine either by an air or liquid cooling system.

Engine efficiency can be discussed in a number of ways but it usually involves a comparison of the total chemical energy in the fuels, and the useful energy extracted from the fuels in the form of kinetic energy. The most fundamental and abstract discussion of engine efficiency is the thermodynamic limit for extracting energy from the fuel defined by a thermodynamic cycle. The most comprehensive is the empirical fuel efficiency of the total engine system for accomplishing a desired task; for example, the miles per gallon accumulated.

Internal combustion engines are primarily heat engines and as such the phenomenon that limits their efficiency is described by thermodynamic cycles. None of these cycles exceed the limit defined by the Carnot cycle which states that the overall efficiency is dictated by the difference between the lower and upper operating temperatures of the engine. A terrestrial engine is usually and fundamentally limited by the upper thermal stability derived from the material used to make up the engine. All metals and alloys eventually melt or decompose and there is significant researching into ceramic materials that can be made with higher thermal stabilities and desirable structural properties. Higher thermal stability allows for greater temperature difference between the lower and upper operating temperatures—thus greater thermodynamic efficiency.

The thermodynamic limits assume that the engine is operating in ideal conditions: a frictionless world, ideal gases, perfect insulators, and operation at infinite time. The real world is substantially more complex and all the complexities reduce the efficiency. In addition, real engines run best at specific loads and rates as described by their power curve. For example, a car cruising on a highway is usually operating significantly below its ideal load, because the engine is designed for the higher loads desired for rapid acceleration. The applications of engines are used as contributed drag on the total system reducing overall efficiency, such as wind resistance designs for vehicles. These and many other losses result in an engines' real-world fuel economy that is usually measured in the units of miles per gallon (or fuel consumption in liters per 100 kilometers) for automobiles. The *miles* in miles per gallon represents a meaningful amount of work and the volume of hydrocarbon implies a standard energy content.

Most steel engines have a thermodynamic limit of 37%. Even when aided with turbochargers and stock efficiency aids, most engines retain an *average* efficiency of about 18%-20%.[7] [8] Rocket engine efficiencies are better still, up to 70%, because they combust at very high temperatures and pressures and are able to have very high expansion ratios.[9]

There are many inventions concerned with increasing the efficiency of IC engines. In general, practical engines are always compromised by trade-offs between different properties such as efficiency, weight, power, heat, response, exhaust emissions, or noise. Sometimes economy also plays a role in not only the cost of manufacturing the engine itself, but also manufacturing and distributing the fuel. Increasing the engines' efficiency brings better fuel economy but only if the fuel cost per energy content is the same.

Measures of fuel/propellant efficiency

For stationary and shaft engines including propeller engines, fuel consumption is measured by calculating the brake specific fuel consumption which measures the number of pounds of fuel that is needed to generate an hours' worth of horsepower-energy. In metric units, the number of grams of fuel needed to generate a kilowatt-hour of energy is calculated.

For internal combustion engines in the form of jet engines, the power output varies drastically with airspeed and a less variable measure is used: thrust specific fuel consumption (TSFC), which is the number of pounds of propellant that is needed to generate impulses that measure a pound an hour. In metric units, the number of grams of propellant needed to generate an impulse that measures one kilonewton per second.

For rockets— TSFC can be used, but typically other equivalent measures are traditionally used, such as specific impulse and effective exhaust velocity.

Air and noise pollution

Internal combustion engines such as reciprocating internal combustion engines produce air pollution emissions, due to incomplete combustion of carbonaceous fuel. The main derivatives of the process are carbon dioxide CO_2, water and some soot—also called particulate matter (PM). The effects of inhaling particulate matter have been studied in humans and animals and include asthma, lung cancer, cardiovascular issues, and premature death. There are however some additional products of the combustion process that include nitrogen oxides and sulfur and some uncombusted hydrocarbons, depending on the operating conditions and the fuel-air ratio.

Not all of the fuel will be completely consumed by the combustion process; a small amount of fuel will be present after combustion, some of which can react to form oxygenates, such as formaldehyde or acetaldehyde, or hydrocarbons not initially present in the fuel mixture. The primary causes of this is the need to operate near the stoichiometric ratio for gasoline engines in order to achieve combustion and the resulting "quench" of the flame by the relatively cool cylinder walls, otherwise the fuel would burn more completely in excess air. When running at lower speeds, quenching is commonly observed in diesel (compression ignition) engines that run on natural gas. It reduces the efficiency and increases knocking, sometimes causing the engine to stall. Increasing the amount of air in the engine reduces the amount of the first two pollutants, but tends to encourage the oxygen and nitrogen in the air to combine to produce nitrogen oxides (NO_x) that has been demonstrated to be hazardous to both plant and animal health. Further chemicals released are benzene and 1,3-butadiene that are also particularly harmful; and not all of the fuel burns up completely, so carbon monoxide (CO) is also produced.

Carbon fuels contain sulfur and impurities that eventually lead to producing sulfur oxides (SO) and sulfur dioxide (SO_2) in the exhaust which promotes acid rain. One final element in exhaust pollution is ozone (O_3). This is not emitted directly but made in the air by the action of sunlight on other pollutants to form "ground level ozone", which, unlike the "ozone layer" in the high atmosphere, is regarded as a bad thing if the levels are too high. Ozone is broken down by nitrogen oxides, so one tends to be lower where the other is higher.

For the pollutants described above (nitrogen oxides, carbon monoxide, sulphur dioxide, and ozone) there are accepted levels that are set by legislation to which no harmful effects are observed—even in sensitive population groups. For the other three: benzene, 1,3-butadiene, and particulates, there is no way of proving they are safe at any level so the experts set standards where the risk to health is, "exceedingly small".

Finally, significant contributions to noise pollution are made by internal combustion engines. Automobile and truck traffic operating on highways and street systems produce noise, as do aircraft flights due to jet noise, particularly supersonic-capable aircraft. Rocket engines create the most intense noise.

See also

- Adiabatic flame temperature
- Air-fuel ratio
- Crude oil engine - a two stroke engine
- Dynamometer
- Electric vehicle
- Engine test stand — information about how to check an internal combustion engine
- External Combustion Engine
- Fossil fuels
- Gas turbine
- Heat pump
- Deglazing (engine mechanics)
- Diesel engine
- Forced induction
- Indirect injection
- Direct injection
- Turbocharger
- Dieselisation
- Gasoline direct injection
- Hybrid vehicle
- Jet engine
- Petrofuel
- Piston engine
- Reciprocating engine
- Steam engine
- IC Engines & EC engines - comparison
- William Barnett — an early patentee (1838)

- variable displacement

Bibliography

- Singer, Charles Joseph; Raper, Richard, *A History of Technology: The Internal Combustion Engine*, edited by Charles Singer ... [et al.], Clarendon Press, 1954-1978. pp. 157–176 [10]
- Hardenberg, Horst O., *The Middle Ages of the Internal combustion Engine*, Society of Automotive Engineers (SAE), 1999

External links

- Animated Engines [11] - explains a variety of types
- Intro to Car Engines [12] - Cut-away images and a good overview of the internal combustion engine
- Walter E. Lay Auto Lab [13] - Research at The University of Michigan
- youtube [14] - Animation of the components and built-up of a 4-cylinder engine
- youtube [15] - Animation of the internal moving parts of a 4-cylinder engine
- Hypervideo showing construction and operation of a four cylinder internal combustion engine courtesy of Ford Motor Company [16]
- Next generation engine technologies [17] retrieved May 9, 2009
- MIT Overview [18] - Present & Future Internal Combustion Engines: Performance, Efficiency, Emissions, and Fuels

References

[1] Encyclopedia Britannica: Internal Combustion engines (http://www.britannica.com/EBchecked/topic/290504/internal-combustion-engine)

[2] Answers.com Internal combustion engine (http://www.answers.com/topic/internal-combustion-engine?cat=technology)

[3] Columbia encyclopedia: Internal combustion engine (http://inventors.about.com/gi/dynamic/offsite.htm?site=http://www.bartleby.com/65/in/intern-co.html)

[4] http://www.infoplease.com/ce6/sci/A0825332.html

[5] (http://www.grc.nasa.gov/WWW/K-12/airplane/otto.html)

[6] Diesel — The Performance Choice (http://www.bankspower.com/Tech_dieselperf.cfm), Banks Talks Tech, 11.19.04

[7] Physics In an Automotive Engine (http://mb-soft.com/public2/engine.html)

[8] Improving IC Engine Efficiency (http://courses.washington.edu/me341/oct22v2.htm)

[9] Rocket propulsion elements 7th edition-George Sutton, Oscar Biblarz pg 37-38

[10] http://proxy.bib.uottawa.ca:2398/cgi/t/text/pageviewer-idx?c=acls&cc=acls&idno=heb02191.0005.001&q1=bicycle&frm=frameset&seq=5

[11] http://www.animatedengines.com

[12] http://www.autoeducation.com/rm_preview/engine_intro.htm

[13] http://me.engin.umich.edu/autolab/Projects/index.html

[14] http://www.youtube.com/watch?v=W2eILCrW53M&NR

[15] http://www.youtube.com/watch?v=2QB7XPMeLnA

[16] http://www.asterpix.com/console/?avi=7660291

[17] http://www.popularmechanics.com/automotive/new_cars/4261289.html?series=19

[18] http://web.mit.edu/professional/short-programs/courses/internal_combustion_engines.html

Extinction

The Dodo, shown here in a 1651 illustration by Jan Savery, is an often-cited example of modern extinction.[1]

In biology and → ecology, **extinction** is the end of an organism or group of taxa. The moment of extinction is generally considered to be the death of the last individual of that species (although the capacity to breed and recover may have been lost before this point). Because a species' potential range may be very large, determining this moment is difficult, and is usually done retrospectively. This difficulty leads to phenomena such as Lazarus taxa, where a species presumed extinct abruptly "re-appears" (typically in the fossil record) after a period of apparent absence.

Through evolution, new species arise through the process of speciation—where new varieties of organisms arise and thrive when they are able to find and exploit an ecological niche—and species become extinct when they are no longer able to survive in changing conditions or against superior competition. A typical species becomes extinct within 10 million years of its first appearance,[2] although some species, called living fossils, survive virtually unchanged for hundreds of millions of years. Extinction, though, is usually a natural phenomenon; it is estimated that 99.9% of all species that have ever lived are now extinct.[2] [3]

Mass extinctions are relatively rare events, however, isolated extinctions are not rare. Starting approximately 100,000 years ago, and coinciding with an increase in the numbers and range of humans, species extinctions have increased to a rate estimated at 100—1000 times that in the recent fossil record.[4] This is known as the Holocene extinction and is at least the sixth such extinction event. Some experts have estimated that up to half of presently existing species may become extinct by 2100.[5]

Definition

External mold of the extinct *Lepidodendron* from the Upper Carboniferous of Ohio.[6]

A species becomes extinct when the last existing member of that species dies. Extinction therefore becomes a certainty when there are no surviving individuals that are able to reproduce and create a new generation. A species may become functionally extinct when only a handful of individuals survive, which are unable to reproduce due to poor health, age, sparse distribution over a large range, a lack of individuals of both sexes (in sexually reproducing species), or other reasons.

Pinpointing the extinction (or pseudoextinction) of a species requires a clear definition of that species. If it is to be declared extinct, the species in question must be uniquely identifiable from any ancestor or daughter species, or from other closely related species. Extinction of a species (or replacement by a daughter species) plays a key role in the punctuated equilibrium hypothesis of Stephen Jay Gould and Niles Eldredge.[7]

In → ecology, *extinction* is often used informally to refer to local extinction, in which a species ceases to exist in the chosen area of study, but still exists elsewhere. This phenomenon is also known as extirpation. Local extinctions may be followed by a replacement of the species taken from other locations; wolf reintroduction is an example of this. Species which are not extinct are termed extant. Those that are extant but threatened by extinction are referred to as threatened or endangered species.

An important aspect of extinction at the present time are human attempts to preserve critically endangered species, which is reflected by the creation of the conservation status "Extinct in the Wild" (EW). Species listed under this status by the World Conservation Union (IUCN) are not known to have any living specimens in the wild, and are maintained only in zoos or other artificial environments. Some of these species are functionally extinct, as they are no longer part of their natural habitat and it is unlikely the species will ever be restored to the wild.[8] When possible, modern zoological institutions attempt to maintain a viable population for species preservation and possible future reintroduction to the wild through use of carefully planned breeding programs.

The extinction of one species' wild population can have knock-on effects, causing further extinctions. These are also called "chains of extinction".[9] This is especially common with extinction of keystone species.

Pseudoextinction

Descendants may or may not exist for extinct species. Daughter species that evolve from a parent species carry on most of the parent species' genetic information, and even though the parent species may become extinct, the daughter species lives on. In other cases, species have produced no new variants, or none that are able to survive the parent species' extinction. Extinction of a parent species where daughter species or subspecies are still alive is also called *pseudoextinction*.

Pseudoextinction is difficult to demonstrate unless one has a strong chain of evidence linking a living species to members of a pre-existing species. For example, it is sometimes claimed that the extinct *Hyracotherium*, which was an early horse that shares a common ancestor with the modern horse, is pseudoextinct, rather than extinct, because there are several extant species of *Equus*, including zebra and donkeys. However, as fossil species typically leave no genetic material behind, it is not possible to say whether *Hyracotherium* actually evolved into more modern horse species or simply evolved from a common ancestor with modern horses. Pseudoextinction is much easier to demonstrate for larger taxonomic groups.

Causes

There are a variety of causes that can contribute directly or indirectly to the extinction of a species or group of species. "Just as each species is unique," write Beverly and Stephen Stearns, "so is each extinction... the causes for each are varied—some subtle and complex, others obvious and simple".[10] Most simply, any species that is unable to survive or reproduce in its environment, and unable to move to a new environment where it can do so, dies out and becomes extinct. Extinction of a species may come suddenly when an otherwise healthy species is wiped out completely, as when toxic pollution renders its entire habitat unlivable; or may occur gradually over thousands or millions of years, such as when a species gradually loses out in competition for food to better adapted competitors.

The Passenger Pigeon, one of hundreds of species of extinct birds, was hunted to extinction over the course of a few decades.

Assessing the relative importance of genetic factors compared to environmental ones as the causes of extinction has been compared to the nature-nurture debate.[3] The question of whether more extinctions in the fossil record have been caused by evolution or by catastrophe is a subject of discussion; Mark Newman,   the   author   of   *Modeling*

Extinction argues for a mathematical model that falls between the two positions.[2] By contrast, conservation biology uses the extinction vortex model to classify extinctions by cause. When concerns about human extinction have been raised, for example in Sir Martin Rees' 2003 book *Our Final Hour*, those concerns lie with the effects of climate change or technological disaster.

Currently, environmental groups and some governments are concerned with the extinction of species caused by humanity, and are attempting to combat further extinctions through a variety of → conservation programs.[4] Humans can cause extinction of a species through overharvesting, pollution, habitat destruction, introduction of new

The Bali Tiger was declared extinct in 1937 due to hunting and habitat loss.

predators and food competitors, overhunting, and other influences. According to the World Conservation Union (WCU, also known as IUCN), 784 extinctions have been recorded since the year 1500 (to the year 2004), the arbitrary date selected to define "modern" extinctions, with many more likely to have gone unnoticed (several species have also been listed as extinct since the 2004 date).[11]

Genetics and demographic phenomena

Population genetics and demographic phenomena affect the evolution, and therefore the risk of extinction, of species. Limited geographic range is the most important determinant of genus extinction at background rates but becomes increasingly irrelevant as mass extinction arises.[12]

Natural selection acts to propagate beneficial genetic traits and eliminate weaknesses. It is nevertheless possible for a deleterious mutation to be spread throughout a population through the effect of genetic drift.

A diverse or deep gene pool gives a population a higher chance of surviving an adverse change in conditions. Effects that cause or reward a loss in genetic diversity can increase the chances of extinction of a species. Population bottlenecks can dramatically reduce genetic diversity by severely limiting the number of reproducing individuals and make inbreeding more frequent. The founder effect can cause rapid, individual-based speciation and is the most dramatic example of a population bottleneck.

Genetic pollution

Purebred, naturally evolved, region specific wild species can be threatened with extinction in a big way[13] through the process of genetic pollution—i.e., uncontrolled hybridization, introgression genetic swamping which leads to homogenization or replacement of local genotypes as a result of a numerical and/or fitness advantage of the introduced plant or animal.[14] Nonnative species can bring about a form of extinction of native plants and animals by hybridization and introgression, either through purposeful introduction by humans or through habitat modification, bringing previously isolated species into contact. These phenomena can be especially detrimental for rare species coming into contact with more abundant ones, where the abundant ones can interbreed with them, swamping the entire rarer gene pool and creating hybrids, thus driving the entire original purebred native stock to complete extinction. Such extinctions are not always apparent from morphological (outward appearance) observations alone. Some degree of gene flow may be a normal, evolutionarily constructive process, and all constellations of genes and genotypes cannot be preserved however, hybridization with or without introgression may, nevertheless, threaten a rare species' existence.[15] [16]

Widespread genetic pollution also leads to weakening of the naturally evolved (wild) region specific gene pool leading to weaker hybrid animals and plants which are not able to cope with natural environs over the long run and fast tracks them towards final extinction.

The gene pool of a species or a population is the complete set of unique alleles that would be found by inspecting the genetic material of every living member of that species or population. A large gene pool indicates extensive genetic diversity, which is associated with robust populations that can survive bouts of intense selection. Meanwhile, low genetic diversity (see inbreeding and population bottlenecks) can cause reduced biological fitness and an increased chance of extinction amongst the reducing population of purebred individuals from a species.

Habitat degradation

The degradation of a species' habitat may alter the fitness landscape to such an extent that the species is no longer able to survive and becomes extinct. This may occur by direct effects, such as the environment becoming toxic, or indirectly, by limiting a species' ability to compete effectively for diminished resources or against new competitor species.

Habitat degradation through toxicity can kill off a species very rapidly, by killing all living members through contamination or sterilizing them. It can also occur over longer periods at lower toxicity levels by affecting life span, reproductive capacity, or competitiveness.

Habitat degradation can also take the form of a physical destruction of niche habitats. The widespread destruction of tropical rainforests and replacement with open pastureland is widely cited as an example of this;[5] elimination of the dense forest eliminated the infrastructure needed by many species to survive. For example, a fern that depends on dense shade for protection from direct sunlight can no longer survive without forest to shelter it. Another example is the destruction of ocean floors by bottom trawling.[17]

Diminished resources or introduction of new competitor species also often accompany habitat degradation. Global warming has allowed some species to expand their range, bringing unwelcome competition to other species that previously occupied that area. Sometimes these new competitors are predators and directly affect prey species, while at other times they may merely outcompete vulnerable species for limited resources. Vital resources including water and food can also be limited during habitat degradation, leading to extinction.

Predation, competition, and disease

Humans have been transporting animals and plants from one part of the world to another for thousands of years, sometimes deliberately (e.g., livestock released by sailors onto islands as a source of food) and sometimes accidentally (e.g., rats escaping from boats). In most cases, such introductions are unsuccessful, but when they do become established as an invasive alien species, the consequences can be catastrophic. Invasive alien species can affect native species directly by eating them, competing with them, and introducing pathogens or parasites that sicken or kill them or, indirectly, by destroying or degrading their habitat. Human populations may themselves act as

The Golden Toad was last seen on May 15, 1989. Decline in amphibian populations is ongoing worldwide.

invasive predators. According to the "overkill hypothesis", the swift extinction of the megafauna in areas such as New Zealand, Australia, Madagascar and Hawaii resulted from the sudden introduction of human beings to environments full of animals that had never seen them before, and were therefore completely unadapted to their predation techniques.[18]

Coextinction

Coextinction refers to the loss of a species due to the extinction of another; for example, the extinction of parasitic insects following the loss of their hosts. Coextinction can also occur when a species loses its pollinator, or to predators in a food chain who lose their prey. "Species coextinction is a manifestation of the interconnectedness of organisms in complex ecosystems ... While coextinction may not be the most important cause of species extinctions, it is certainly an insidious one".[19] Coextinction is especially common when a keystone species goes extinct.

Global warming

There is also discussion about the long term affects of global warming on the extinction process. Currently, studies have concluded that global warming may drive one quarter of all land animals and plants to extinction by 2050.[20] The absolute worst case scenario that we are facing is a thrilling 1/3 to 1/2 of all plant and animal species facing extinction.[21] The ecologically rich hot spots where potentially most damage would be done include places like South Africa's Cape Floristic Region, and the Caribbean Basin. These areas include a doubling of present carbon dioxide levels and rising temperatures that could eliminate 56,000 plant and 3,700 animal species in these hot spot regions.[22]

The white form of the lemuroid possum, only found in the mountain forests of northern Queensland, was once named as the first mammal species sub-form to be driven extinct by global warming.[23] However since then 3 possums have been found.[24] Also a more common brown form of the lemuroid possum is only considered "near threatened" and is not at risk of extinction.[25] Climate change however is only one threat to animal and plant species. Plants and animals are also feeling the devastating effects of deforestation and habitat destruction.[26]

Mass extinctions

There have been at least five mass extinctions in the history of life on earth, and four in the last 3.5 billion years in which many species have disappeared in a relatively short period of geological time. The most recent of these, the Cretaceous–Tertiary extinction event 65 million years ago at the end of the Cretaceous period, is best known for having wiped out the non-avian dinosaurs, among many other species.

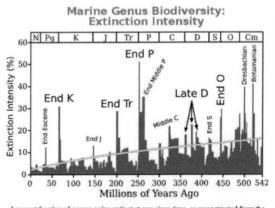

Apparent fraction of genera going extinct at any given time, as reconstructed from the fossil record.

Modern extinctions

According to a 1998 survey of 400 biologists conducted by New York's American Museum of Natural History, nearly 70 percent believed that they were currently in the early stages of a human-caused extinction,[27] known as the Holocene extinction. In that survey, the same proportion of respondents agreed with the prediction that up to 20 percent of all living populations could become extinct within 30 years (by 2028). Biologist E. O. Wilson estimated [5] in 2002 that if current rates of human destruction of the biosphere continue, one-half of all species of life on earth will be extinct in 100 years.[28] More significantly the rate of species extinctions at present is estimated at 100 to 1000 times "background" or average extinction rates in the evolutionary time scale of planet Earth.[29]

History of scientific understanding

In the 1800s when extinction was first described, the idea of extinction was threatening to those who held a belief in the Great Chain of Being, a theological position that did not allow for "missing links".[30]

The possibility of extinction was not widely accepted before the 1800s.[30] [31] The devoted naturalist Carl Linnaeus, could "hardly entertain" the idea that humans could cause the extinction of a species.[32] When parts of the world had not been thoroughly examined and charted, scientists could not rule out that animals found only in the fossil record were not simply "hiding" in unexplored regions of the Earth.[33] Georges Cuvier is credited with establishing extinction as a fact in a 1796 lecture to the French Institute.[31] Cuvier's observations of fossil bones convinced him that they did not originate in extant animals. This discovery was critical for the spread of uniformitarianism,[34] and lead to the first book publicizing the idea of evolution [35] though Cuvier himself strongly opposed the theories of evolution advanced by Lamarck and others.

Dilophosaurus, one of the many extinct dinosaur genera. The cause of the Cretaceous–Tertiary extinction event is a subject of much debate amongst researchers.

Human attitudes and interests

Extinction is an important research topic in the field of zoology, and biology in general, and has also become an area of concern outside the scientific community. A number of organizations, such as the Worldwide Fund for Nature, have been created with the goal of preserving species from extinction. Governments have attempted, through enacting laws, to avoid habitat destruction, agricultural over-harvesting, and pollution. While many human-caused extinctions have been accidental, humans have also engaged in the deliberate destruction of some species, such as dangerous viruses, and the total destruction of other problematic species has been suggested. Other species were deliberately driven to extinction, or nearly so, due to poaching or because they were "undesirable", or to push for other human agendas. One example was the near extinction of the American bison, which was nearly wiped out by mass hunts sanctioned by the United States government, in order to force the removal of Native Americans, many of whom relied on the bison for food.

Biologist Bruce Walsh of the University of Arizona states three reasons for scientific interest in the preservation of species; genetic resources, ecosystem stability, and ethics;[36] and today the scientific community "stress[es] the importance" of maintaining biodiversity.[36] [37]

In modern times, commercial and industrial interests often have to contend with the effects of production on plant and animal life. However, some technologies with minimal, or no, proven harmful effects on *Homo sapiens* can be devastating to wildlife (for example, DDT).[38] Biogeographer Jared Diamond notes that while big business may label environmental concerns as "exaggerated", and often cause "devastating damage", some corporations find it in their interest to adopt good conservation practices, and even engage in preservation efforts that surpass those taken by national parks.[39]

Governments sometimes see the loss of native species as a loss to ecotourism,[40] and can enact laws with severe punishment against the trade in native species in an effort to prevent extinction in the wild. Nature preserves are created by governments as a means to provide continuing habitats to species crowded by human expansion. The 1992 Convention on Biological Diversity has resulted in international Biodiversity Action Plan programmes, which attempt to provide comprehensive guidelines for government biodiversity conservation. Advocacy groups, such as The Wildlands Project[41] and the Alliance for Zero Extinctions,[42] work to educate the public and pressure governments into action.

People who live close to nature can be dependent on the survival of all the species in their environment, leaving them highly exposed to extinction risks. However, people prioritize day-to-day survival over species conservation; with human overpopulation in tropical developing countries, there has been enormous pressure on forests due to subsistence agriculture, including slash-and-burn agricultural techniques that can reduce endangered species's habitats.[43]

Planned extinction

Humans have aggressively worked toward the extinction of many species of viruses and bacteria in the cause of disease eradication. For example, the smallpox virus is now extinct in the wild[44] —although samples are retained in laboratory settings, and the polio virus is now confined to small parts of the world as a result of human efforts to prevent the disease it causes.[45]

Olivia Judson is one of six modern scientists to have advocated the deliberate extinction of specific species. Her September 25, 2003 *New York Times* article, "A Bug's Death", advocates "specicide" of thirty mosquito species through the introduction of a genetic element, capable of inserting itself into another crucial gene, to create recessive "knockout genes". Her arguments for doing so are that the *Anopheles* mosquitoes (which spread malaria) and *Aedes* mosquitoes (which spread dengue fever, yellow fever, elephantiasis, and other diseases) represent only 30 species; eradicating these would save at least one million human lives per annum at a cost of reducing the genetic diversity of the family Culicidae by only 1%. She further argues that since species become extinct "all the time" the disappearance of a few more will not destroy the ecosystem: "We're not left with a wasteland every time a species vanishes. Removing one species sometimes causes shifts in the populations of other species — but different need not mean worse." In addition, anti-malarial and mosquito control programs offer little realistic hope to the 300 million people in developing nations who will be infected with acute illnesses this year. Although trials are ongoing, she writes that if they fail: "We should consider the ultimate swatting."[46]

Cloning

Recent technological advances have encouraged the hypothesis that by using DNA from the remains of an extinct species, through the process of cloning, the species may be "brought back to life". Proposed targets for cloning include the mammoth, thylacine, and the Pyrenean Ibex. In order for such a program to succeed, a sufficient number of individuals would have to be cloned, from the DNA of different individuals (in the case of sexually reproducing organisms) to create a viable population. Though bioethical and philosophical objections have been raised, the cloning of extinct creatures seems a viable outcome of the continuing advancements in our science and technology.

In 2003, scientists attempted to clone the extinct Pyrenean Ibex (C. p. pyrenaica). This initial attempt failed; of the 285 embryos reconstructed, 54 were transferred to 12 mountain goats and mountain goat-domesticated goat hybrids, but only two survived the initial two months of gestation before they too died.

In 2009, a second attempt was made to clone the Pyrenean Ibex; one clone was born alive, but died seven minutes later, due to physical defects in the lungs.[47]

The concept of cloning extinct species was thought to be popularized by the successful novel and movie Jurassic Park, though it may have been first used in John Brosnan's 1984 novel Carnosaur, and later in Piers Anthony's Balook novel, which resurrected a Baluchitherium.

See also

- Gene pool
- Genetic erosion
- Genetic pollution
- Habitat fragmentation
- IUCN Red List
- List of extinct animals
- List of extinct plants
- Living Planet Index
- Mass extinction
- Red List Index
- Refugium (population biology)
- Timeline of extinctions
- Voluntary Human Extinction Movement

External links

- Committee on recently extinct organisms [48]
- Recently Extinct Animals [49]
- H.E. Strickland's *The Dodo and its Kindred* (London: 1848) [50], a study of three extinct bird species — full digital facsimile, Linda Hall Library
- Sepkoski's Global Genus Database of Marine Animals [51] — Calculate past mass extinction rates for yourself!

References

[1] Diamond, Jared (1999). "Up to the Starting Line". *Guns, Germs, and Steel*. W. W. Norton. pp. 43–44. ISBN 0-393-31755-2.

[2] Newman, Mark. " A Mathematical Model for Mass Extinction (http://www.lassp.cornell.edu/newmme/science/extinction.html)". Cornell University. May 20, 1994. URL accessed July 30, 2006.

[3] Raup, David M. *Extinction: Bad Genes or Bad Luck?* W.W. Norton and Company. New York. 1991. pp.3-6 ISBN 978-0393309270

[4] Species disappearing at an alarming rate, report says (http://www.msnbc.msn.com/id/6502368/). MSNBC URL accessed July 26, 2006

[5] Wilson, E.O., *The Future of Life* (2002) (ISBN 0-679-76811-4). See also: Leakey, Richard, *The Sixth Extinction : Patterns of Life and the Future of Humankind*, ISBN 0-385-46809-1

[6] Davis, Paul and Kenrick, Paul. Fossil Plants. Smithsonian Books, Washington D.C. (2004). Morran, Robin, C.; A Natural History of Ferns. Timber Press (2004). ISBN 0-88192-667-1

[7] See: Niles Eldredge, *Time Frames: Rethinking of Darwinian Evolution and the Theory of Punctuated Equilibria*, 1986, Heinemann ISBN 0-434-22610-6

[8] Maas, Peter. "[http://www.petermaas.nl/extinct/wilduk.htm Extinct in the Wild" *The Extinction Website*. URL accessed January 26 2007.

[9] Quince, C. et al. (pdf). *Deleting species from model food webs* (http://theory.ph.man.ac.uk/~ajm/qui05a.pdf). . Retrieved 2007-02-15.

[10] Stearns, Beverly Peterson and Stephen C. (2000). "Preface". *Watching, from the Edge of Extinction*. Yale University Press. pp. x. ISBN 0300084692.

[11] " 2004 Red List (http://web.archive.org/web/20080212161231/http://www.iucn.org/themes/ssc/red_list_2004/GSAexecsumm_EN. htm)". *IUCN Red List of Threatened Species*. World Conservation Union. Archived from the original (http://www.iucn.org/themes/ssc/red_list_2004/GSAexecsumm_EN.htm) on 12 February 2008. . Retrieved September 20, 2006.

[12] Payne, J.L. & S. Finnegan (2007). "The effect of geographical range on extinction risk during background and mass extinction.". *Proc. Nat. Acad. Sci.* **104** (25): 10506–11. doi: 10.1073/pnas.0701257104 (http://dx.doi.org/10.1073/pnas.0701257104). PMID 17563357 (http://www.ncbi.nlm.nih.gov/pubmed/17563357).

[13] Mooney, H. A.; Cleland, E. E. (2001). "The evolutionary impact of invasive species". *PNAS* **98** (10): 5446–5451. doi: 10.1073/pnas.091093398 (http://dx.doi.org/10.1073/pnas.091093398).

[14] Glossary: definitions from the following publication: Aubry, C., R. Shoal and V. Erickson. 2005. Grass cultivars: their origins, development, and use on national forests and grasslands in the Pacific Northwest. USDA Forest Service. 44 pages, plus appendices.; Native Seed Network (NSN), Institute for Applied Ecology, 563 SW Jefferson Ave, Corvallis, OR 97333, USA (http://www.nativeseednetwork.org/article_view?id=13)

[15] Extinction by Hybridization and Introgression; by Judith M. Rhymer , Department of Wildlife Ecology, University of Maine, Orono, Maine 04469, USA; and Daniel Simberloff, Department of Biological Science, Florida State University, Tallahassee, Florida 32306, USA; Annual

Review of Ecology and Systematics, November 1996, Vol. 27, Pages 83-109 (doi: 10.1146/annurev.ecolsys.27.1.83) (http://arjournals. annualreviews.org/doi/abs/10.1146/annurev.ecolsys.27.1.83), (http://links.jstor.org/sici?sici=0066-4162(1996)27<83:EBHAI>2.0. CO;2-A#abstract)

[16] Genetic Pollution from Farm Forestry using eucalypt species and hybrids; A report for the RIRDC/L&WA/FWPRDC; Joint Venture Agroforestry Program; by Brad M. Potts, Robert C. Barbour, Andrew B. Hingston; September 2001; RIRDC Publication No 01/114; RIRDC Project No CPF - 3A; ISBN 0642583366; ISSN 1440-6845; Australian Government, Rural Industrial Research and Development Corporation (http://www.rirdc.gov.au/reports/AFT/01-114.pdf)

[17] Clover, Charles. 2004. *The End of the Line: How overfishing is changing the world and what we eat.* Ebury Press, London. ISBN 0-09-189780-7

[18] Lee, Anita. " The Pleistocene Overkill Hypothesis (http://geography.berkeley.edu/ProgramCourses/CoursePagesFA2002/geog148/ Term Papers/Anita Lee/THEPLE~1.html)." *University of California at Berkeley Geography Program.* URL accessed January 11, 2007.

[19] Koh, Lian Pih. *Science*, Vol 305, Issue 5690, 1632-1634, 10 September 2004.

[20] Battachatya, Shaoni. (http://www.newscientist.com/article/dn4545-global-warming-threatens-millions-of-species.html)." URL accessed September 15, 2008.

[21] Bhattacharya, Shaoni. "Global warming threatens millions of species." Global Warming (Jan. 2004): n. pag. www.newscientist.com. Web. 12 Oct. 2009. <http://www.newscientist.com/article/ dn4545-global-warming-threatens-millions-of-species.html>.

[22] Handwerk, Brian, and Brian Hendwerk. "Global Warming Could Cause Mass Extinctions by 2050, Study Says." National Geographic News (Apr. 2006): n. pag. www.nationalgeographic.com. Web. 12 Oct. 2009.

[23] White possum said to be first victim of global warming (http://www.news.com.au/couriermail/story/0,23739,24742053-952,00.html)

[24] (http://news.ninemsn.com.au/national/794434/extinct-possum-found-in-daintree)

[25] Burnett, S. & Winter, J. (2008). *Hemibelideus lemuroides* (http://www.iucnredlist.org/apps/redlist/details/9869). In: IUCN 2008. IUCN Red List of Threatened Species. Downloaded on 28 December 2008. Database entry includes justification for why this species is listed as near threatened

[26] Handwerk, Brian. "Global Warming Could Cause Mass Extinctions by 2050, Study Says." National Geographic News (Apr. 2006): n. pag. www.nationalgeographic.com. Web. 12 Oct. 2009.

<http://news.nationalgeographic.com/news/2006/04/ 0412_060412_global_warming.html>.

[27] American Museum of Natural History. " National Survey Reveals Biodiversity Crisis - Scientific Experts Believe We are in the Midst of the Fastest Mass Extinction in Earth's History (http://www.well.com/~davidu/amnh.html)". URL accessed September 20, 2006.

[28] Ulansey, David, " The current mass extinction (http://www.well.com/user/davidu/extinction.html)" repeats this statement with links to dozens of news reports on the phenomenon. URL accessed January 26, 2007.

[29] J.H.Lawton and R.M.May, *Extinction rates*, Oxford University Press, Oxford, UK

[30] Viney, Mike. " Extinction Part 2 of 5 (http://www.csmate.colostate.edu/cltw/cohortpages/viney_old1/extinction.html)". Colorado State University. URL accessed September 12, 2006.

[31] Academy of Natural Sciences, "Fossils and Extinction" (http://www.ansp.org/museum/jefferson/otherPages/extinction.php) and U.C. Berkeley "History of Evolutionary Thought — Extinction" http://evolution.berkeley.edu/evosite/history/extinction.shtml.

[32] Koerner, Lisbet (1999). "God's Endless Larder". *Linnaeus: Nature and Nation*. Harvard University Press. pp. 85. ISBN 0-674-00565-1.

[33] *Ideas: A History from Fire to Freud* (Peter Watson Weidenfeld & Nicolson ISBN 0-297-60726-X)

[34] Watson, p.16

[35] Robert Chambers, 1844, *Vestiges of the Natural History of Creation*, 1994 reprint: University of Chicago Press ISBN 0-226-10073-1

[36] Walsh, Bruce. Extinction (http://nitro.biosci.arizona.edu/courses/EEB105/lectures/extinction/extinction.html). Bioscience at University of Arizona. URL accessed July 26, 2006.

[37] Committee on Recently Extinct Organisms. " Why Care About Species That Have Gone Extinct? (http://creo.amnh.org/care.html)". URL accessed July 30, 2006.

[38] International Programme on Chemical Safety (1989). " DDT and its Derivatives -- Environmental Aspects (http://www.inchem.org/ documents/ehc/ehc/ehc83.htm)". Environmental Health Criteria 83. URL accessed September 20, 2006.

[39] Diamond, Jared (2005). "A Tale of Two Farms". *Collapse*. Penguin. pp. 15–17. ISBN 0-670-03337-5.

[40] Drewry, Rachel. " Ecotourism: Can it save the orangutans? (http://www.insideindonesia.org/edit51/orang.htm)" *Inside Indonesia*. URL accessed January 26, 2007.

[41] The Wildlands Project (http://www.wild-earth.org/cms/page1090.cfm). URL accessed January 26, 2007.

[42] Alliance for Zero Extinctions (http://www.zeroextinction.org/). URL accessed January 26, 2007.

[43] Ehrlich, Anne (1981). *Extinction: The Causes and Consequences of the Disappearance of Species*. Random House, New York. ISBN 0-394-51312-6.

[44] WHO Factsheet (http://www.who.int/mediacentre/factsheets/smallpox/) WHO meeting agenda (http://ftp.who.int/gb/pdf_files/ WHA52/ew5.pdf) Scientists certified it eradicated in December 1979, WHO formally ratified this on 8 May 1980 in resolution WHA33.3

[45] Global Polio Eradication Initiative. " The History (http://www.polioeradication.org/history.asp)". URL accessed January 24, 2007.

[46] Judson, Olivia (September 25, 2003). " "A Bug's Death" (http://query.nytimes.com/gst/fullpage.html?sec=health& res=9805E5DF143DF936A1575AC0A9659C8B63&n=Top/News/Science/Topics/Mosquitoes)". *New York Times*. . Retrieved 2006-07-30.

[47] (http://www.telegraph.co.uk/science/science-news/4409958/Extinct-ibex-is-resurrected-by-cloning.html)

[48] http://creo.amnh.org/

[49] http://extinctanimals.petermaas.nl/
[50] http://contentdm.lindahall.org/u?/Natural_His,2106
[51] http://strata.geology.wisc.edu/jack

Rachel Carson

Rachel Louise Carson

Rachel Carson, 1940
Fish & Wildlife Service employee photo

Born	May 27, 1907 Springdale, Pennsylvania
Died	April 14, 1964 (aged 56) Silver Spring, Maryland
Occupation	marine biologist, writer
Nationality	American
Writing period	1937–1964
Genres	nature writing
Subjects	marine biology, → ecology, pesticides
Notable work(s)	*Silent Spring*

Rachel Louise Carson (May 27, 1907 – April 14, 1964) was an American marine biologist and nature writer whose writings are credited with advancing the global → environmental movement.

Carson started her career as a biologist in the U.S. Bureau of Fisheries, and became a full-time nature writer in the 1950s. Her widely praised 1951 bestseller *The Sea Around Us* won her financial security and recognition as a gifted writer. Her next book, *The Edge of the Sea*, and the republished version of her first book, *Under the Sea Wind*, were also bestsellers. Together, her sea trilogy explores the whole of ocean life, from the shores to the surface to the deep sea.

In the late 1950s, Carson turned her attention to conservation and the environmental problems caused by synthetic pesticides. The result was *Silent Spring* (1962), which brought environmental concerns to an unprecedented portion of the American public. *Silent Spring* spurred a reversal in national pesticide policy—leading to a nationwide ban on

DDT and other pesticides—and the grassroots environmental movement the book inspired led to the creation of the Environmental Protection Agency. Carson was posthumously awarded the Presidential Medal of Freedom by Jimmy Carter.

Life and work

Carson's childhood home now is preserved as the Rachel Carson Homestead

Early life and education

Rachel Carson was born on May 27, 1907 on a small family farm near Springdale, Pennsylvania, just up the Allegheny River from Pittsburgh. Carson was an avid reader, and, from a remarkably young age, a talented writer. She also spent a lot of time exploring around her 65-acre (260000 m^2) farm. She began writing stories (often involving animals) at age eight, and had her first story published at age eleven. She especially enjoyed the *St. Nicholas Magazine* (which carried her first published stories),the works of Beatrix Potter, and the novels of Gene Stratton Porter, and in her teen years, Herman Melville, Joseph Conrad and Robert Louis Stevenson. The natural world, particularly the ocean, was the common thread of her favorite literature. Carson attended Springdale's small school through tenth grade, then completed high school in nearby Parnassus, Pennsylvania, graduating in 1925 at the top of her class of forty-four students.[1]

At the Pennsylvania College for Women (today known as Chatham University), as in high school, Carson was somewhat of a loner. She originally studied English, but switched her major to biology in January 1928, though she continued contributing to the school's student newspaper and literary supplement. Though admitted to graduate standing at Johns Hopkins University in 1928, she was forced to remain at the Pennsylvania College for Women for her senior year due to financial difficulties; she graduated *magna cum laude* in 1929. After a summer course at the Marine Biological Laboratory, she continued her studies in zoology and genetics at Johns Hopkins in the fall of 1929.[2]

After her first year of graduate school, Carson became a part-time student, taking an assistantship in Raymond Pearl's laboratory, where she worked with rats and *Drosophila*, to earn money for tuition. After false starts with pit vipers and squirrels, she completed a dissertation project on the embryonic development of the pronephros in fish. She earned a master's degree in zoology in June 1932. She had intended to continue for a doctorate, but in 1934 Carson was forced to leave Johns Hopkins to search for a full-time teaching position to help support her family. In 1935, her father died suddenly, leaving Carson to care for her aging mother and making the financial situation even more critical. At the urging of her undergraduate biology mentor Mary Scott Skinker, she settled for a temporary position with the U.S. Bureau of Fisheries writing radio copy for a series of weekly educational broadcasts entitled "Romance Under the Waters". The series of fifty-two seven-minute programs focused on aquatic life and was intended to generate public interest in fish biology and in the work of the bureau—a task the several writers before Carson had not managed. Carson also began submitting articles on marine life in the Chesapeake Bay, based on her research for the series, to local newspapers and magazines.[3]

Carson's supervisor, pleased with the success of the radio series, asked her to write the introduction to a public brochure about the fisheries bureau; he also worked to secure her the first full-time position that became available. Sitting for the civil service exam, she outscored all other applicants and in 1936 became only the second woman to be hired by the Bureau of Fisheries for a full-time, professional position, as a junior aquatic biologist.[4]

Early career and publications

At the U.S. Bureau of Fisheries, Carson's main responsibilities were to analyze and report field data on fish populations, and to write brochures and other literature for the public. Using her research and consultations with marine biologists as starting points, she also wrote a steady stream of articles for *The Baltimore Sun* and other newspapers. However, her family responsibilities further increased in January 1937 when her older sister died, leaving Carson as the sole breadwinner for her mother and two nieces.[5]

In July 1937, the *Atlantic Monthly* accepted a revised version of an essay, "The World of Waters", that she had originally written for her first fisheries bureau brochure; her supervisor had deemed it too good for that purpose. The essay, published as "Undersea", was a vivid narrative of a journey along the ocean floor. It marked a major turning point in Carson's writing career. Publishing house Simon & Schuster, impressed by "Undersea", contacted Carson and suggested that she expand it into book form. Several years of writing resulted in *Under the Sea Wind* (1941), which received excellent reviews but sold poorly. In the meantime, Carson's article-writing success continued—her features appeared in *Sun Magazine, Nature*, and *Collier's*.[6]

Carson attempted to leave the Bureau (by then transformed into the Fish and Wildlife Service) in 1945, but few jobs for naturalists were available as most money for science was focused on technical fields in the wake of the Manhattan Project. In mid-1945, Carson first encountered the subject of DDT, a revolutionary new pesticide (lauded as the "insect bomb" after the atomic bombings of Hiroshima and Nagasaki) that was only beginning to undergo tests for safety and ecological effects. DDT was but one of Carson's many writing interests at the time, and editors found the subject unappealing; she published nothing on DDT until 1962.[7]

Carson rose within the Fish and Wildlife Service, supervising a small writing staff by 1945 and becoming chief editor of publications in 1949. Though her position provided increasing opportunities for fieldwork and freedom in choosing her writing projects, it also entailed increasingly tedious administrative responsibilities. By 1948, Carson was working on material for a second book and had made the conscious decision to begin a transition to writing full-time. That year, she took on a literary agent, Marie Rodell; they formed a close professional relationship that would last the rest of Carson's career.[8]

Oxford University Press expressed interest in Carson's book proposal for a life history of the ocean, spurring her to complete the manuscript of what would become *The Sea Around Us* by early 1950.[9] Chapters appeared in *Science Digest* and the *Yale Review*—the latter chapter, "The Birth of an Island", winning the American Association for the Advancement of Science's George Westinghouse Science Writing Prize—and nine chapters were serialized in *The New Yorker*. *The Sea Around Us* remained on the *New York Times* bestseller list for 86 weeks, was abridged by *Reader's Digest*, won the 1952 National Book Award and the Burroughs Medal, and resulted in Carson being awarded two honorary doctorates. She also licensed a documentary film to be based on *The Sea Around Us*. The book's success led to the republication of *Under the Sea Wind*, which also became a best-seller. With success came financial security, and Carson was able to give up her job in 1952 to concentrate on writing full time.[10]

Carson was inundated with speaking engagements, fan mail and other correspondence regarding *The Sea Around Us*, along with work on the documentary script that she had secured the right to review.[11] She was extremely unhappy with the final version of the script by writer, director and producer Irwin Allen; she found it untrue to the atmosphere of the book and scientifically embarrassing, describing it as "a cross between a believe-it-or-not and a breezy travelogue."[12] She discovered, however, that her right to review the script did not extend to any control over its content. Allen proceeded in spite of Carson's objections to produce a very successful documentary. It won the 1953 Oscar for Best Documentary, but Carson was so embittered by the experience that she never again sold film rights to her work.[13]

Relationship with Dorothy Freeman

Carson moved with her mother to Southport Island, Maine in 1953, and in July of that year met Dorothy Freeman (1898–1978)—the beginning of an extremely close relationship that would last the rest of Carson's life. The nature of the relationship between Carson and Freeman has been the subject of much interest and speculation. It is probably best described as a romantic friendship. Carson met Freeman, a summer resident of the island along with her husband, after Freeman had written to Carson to welcome her. Freeman had read *The Sea Around Us*, a gift from her son, and was excited to have the prominent author as a neighbor. Carson's biographer Linda Lear writes that "Carson sorely needed a devoted friend and kindred spirit who would listen to her without advising and accept her wholly, the writer as well as the woman."[14] She found this in Freeman. The two women had a number of common interests, nature chief among them, and began exchanging letters regularly while apart. They would continue to share every summer for the remainder of Carson's life, and meet whenever else their schedules permitted.[15]

Though Lear does not explicitly describe the relationship as romantic, others (such as the encyclopedia *glbtq*[16]) have noted that Carson and Freeman knew that their letters could be interpreted as lesbian, even though "the expression of their love was limited almost wholly to letters and very occasional farewell kisses or holding of hands."[17] Freeman shared parts of Carson's letters with her husband to help him understand the relationship, but much of their correspondence was carefully guarded.[18] Shortly before Carson's death, she and Freeman destroyed hundreds of letters. The surviving correspondence was published in 1995 as *Always, Rachel: The Letters of Rachel Carson and Dorothy Freeman, 1952–1964: An Intimate Portrait of a Remarkable Friendship*, edited by Freeman's granddaughter. According to one reviewer, the pair "fit Carolyn Heilbrun's characterization of a strong female friendship, where what matters is 'not whether friends are homosexual or heterosexual, lovers or not, but whether they share the wonderful energy of work in the public sphere'".[19]

The Edge of the Sea and transition to conservation work

In early 1953 Carson began library and field research on the ecology and organisms of the Atlantic shore.[20] In 1955, she completed the third volume of her sea trilogy, *The Edge of the Sea*, which focuses on life in coastal ecosystems (particularly along the Eastern Seaboard). It appeared in *The New Yorker* in two condensed installments shortly before the October 26 book release. By this time, Carson's reputation for clear and poetical prose was well-established; *The Edge of the Sea* received highly favorable reviews, if not quite as enthusiastic as for *The Sea Around Us*.[21]

Through 1955 and 1956, Carson worked on a number of projects—including the script for an *Omnibus* episode, "Something About the Sky"—and wrote articles for popular magazines. Her plan for the next book was to address evolution, but the publication of Julian Huxley's *Evolution in Action*—and her own difficulty in finding a clear and compelling approach to the topic—led her to abandon the project. Instead, her interests were turning to conservation. She considered an environment-themed book project tentatively entitled *Remembrance of the Earth* and became involved with The Nature Conservancy and other conservation groups. She also made plans to buy and preserve from development an area in Maine she and Freeman called the "Lost Woods".[22]

Early in 1957, family tragedy struck a third time when one of the nieces she had cared for in the 1940s died at the age of 31, leaving a five-year-old orphan son, Roger Christie. Carson took on that responsibility, adopting the boy, alongside continuing to care for her aging mother; this took a considerable toll on Carson. She moved to Silver Spring, Maryland to care for Roger, and much of 1957 was spent putting their new living situation in order and focusing on specific environmental threats.[23]

By fall 1957, Carson was closely following federal proposals for widespread pesticide spraying; the USDA planned to eradicate fire ants, and other spraying programs involving chlorinated hydrocarbons and organophosphates were on the rise.[24] For the rest of her life, Carson's main professional focus would be the dangers of pesticide overuse.

Silent Spring

Research and writing

Starting in the mid-1940s, Carson had become concerned about the use of synthetic pesticides, many of which had been developed through the military funding of science since World War II. It was the USDA's 1957 fire ant eradication program, however, that prompted Carson to devote her research, and her next book, to pesticides and environmental poisons. The fire ant program involved aerial spraying of DDT and other pesticides (mixed with fuel oil), including the spraying of private land. Landowners in Long Island filed a suit to have the spraying stopped, and many in affected regions followed the case closely. Though the suit was lost, the Supreme Court granted petitioners the right to gain injunctions against potential environmental damage in the future; this laid the basis for later successful environmental actions.[25]

The Washington, D.C. chapter of the Audubon Society also actively opposed such spraying programs, and recruited Carson to help make public the government's exact spraying practices and the related research.[26] Carson began the four-year project of what would become *Silent Spring* by gathering examples of environmental damage attributed to DDT. She also attempted to enlist others to join the cause: essayist E. B. White, and a number of journalists and scientists. By 1958, Carson had arranged a book deal, with plans to co-write with *Newsweek* science journalist Edwin Diamond. However, when *The New Yorker* commissioned a long and well-paid article on the topic from Carson, she began considering writing more than simply the introduction and conclusion as planned; soon it was a solo project. (Diamond would later write one of the harshest critiques of *Silent Spring*.)[27]

As her research progressed, Carson found a sizable community of scientists who were documenting the physiological and environmental effects of pesticides. She also took advantage of her personal connections with many government scientists, who supplied her with confidential information. From reading the scientific literature and interviewing scientists, Carson found two scientific camps when it came to pesticides: those who dismissed the possible danger of pesticide spraying barring conclusive proof, and those who were open to the possibility of harm and willing to consider alternative methods such as biological pest control.[28]

By 1959, the USDA's Agricultural Research Service responded to the criticism of Carson and others with a public service film, *Fire Ants on Trial*; Carson characterized it as "flagrant propaganda" that ignored the dangers that spraying pesticides (especially dieldrin and heptachlor) posed to humans and wildlife. That spring, Carson wrote a letter, published in *The Washington Post*, that attributed the recent decline in bird populations—in her words, the "silencing of birds"—to pesticide overuse.[29] That was also the year of the "Great Cranberry Scandal": the 1957, 1958, and 1959 crops of U.S. cranberries were found to contain high levels of the herbicide aminotriazole (which caused cancer in laboratory rats) and the sale of all cranberry products was halted. Carson attended the ensuing FDA hearings on revising pesticide regulations; she came away discouraged by the aggressive tactics of the chemical industry representatives, which included expert testimony that was firmly contradicted by the bulk of the scientific literature she had been studying. She also wondered about the possible "financial inducements behind certain pesticide programs".[30]

Research at the Library of Medicine of the National Institutes of Health brought Carson into contact with medical researchers investigating the gamut of cancer-causing chemicals. Of particular significance was the work of National Cancer Institute researcher and founding director of the environmental cancer section Wilhelm Hueper, who classified many pesticides as carcinogens. Carson and her research assistant Jeanne Davis, with the help of NIH librarian Dorothy Algire, found evidence to support the pesticide-cancer connection; to Carson the evidence for the toxicity of a wide array of synthetic pesticides was clear-cut, though such conclusions were very controversial beyond the small community of scientists studying pesticide carcinogenesis.[31]

By 1960, Carson had more than enough research material, and the writing was progressing rapidly. In addition to the thorough literature search, she had investigated hundreds of individual incidents of pesticide exposure and the human sickness and ecological damage that resulted. However, in January, a duodenal ulcer followed by several infections kept her bedridden for weeks, greatly delaying the completion of *Silent Spring*. As she was nearing full recovery in March (just as she was completing drafts of the two cancer chapters of her book), she discovered cysts in her left breast, one of which necessitated a mastectomy. Though her doctor described the procedure as precautionary and recommended no further treatment, by December Carson discovered that the tumor was in fact malignant and the cancer had metastasized.[32] Her research was also delayed by revision work for a new edition of *The Sea Around Us*, and by a collaborative photo essay with Erich Hartmann.[33] Most of the research and writing was done by the fall of 1960, except for the discussion of recent research on biological controls and investigations of a handful of new pesticides. However, further health troubles slowed the final revisions in 1961 and early 1962.[34]

It was difficult finding a title for the book; "Silent Spring" was initially suggested as a title for the chapter on birds. By August 1961, Carson finally agreed to the suggestion of her literary agent Marie Rodell: *Silent Spring* would be a metaphorical title for the entire book—suggesting a bleak future for the whole natural world—rather than a literal chapter title about the absence of birdsong.[35] With Carson's approval, editor Paul Brooks at Houghton Mifflin arranged for illustrations by Louis and Lois Darling, who also designed the cover. The final writing was the first chapter, "A Fable for Tomorrow", which was intended to provide a gentler introduction to what might otherwise be a forbiddingly serious topic. By mid-1962, Brooks and Carson had largely finished the editing, and were laying the groundwork for promoting the book by sending the manuscript out to select individuals for final suggestions.[36]

Argument

As biographer Mark Hamilton Lytle writes, Carson "quite self-consciously decided to write a book calling into question the paradigm of scientific progress that defined postwar American culture." The overriding theme of *Silent Spring* is the powerful—and often negative—effect humans have on the natural world.[37]

Carson's main argument is that pesticides have detrimental effects on the environment; they are more properly termed "biocides", she argues, because their effects are rarely limited to the target pests. DDT is a prime example, but other synthetic pesticides come under scrutiny as well—many of which are subject to bioaccumulation. Carson also accuses the chemical industry of intentionally spreading disinformation and public officials of accepting industry claims uncritically. Most of the book is devoted to pesticides' effects on natural ecosystems, but four chapters also detail cases of human pesticide poisoning, cancer, and other illnesses attributed to pesticides.[38] About DDT and cancer, the subject of so much subsequent debate, Carson says only a little:

> In laboratory tests on animal subjects, DDT has produced suspicious liver tumors. Scientists of the Food and Drug Administration who reported the discovery of these tumors were uncertain how to classify them, but felt there was some "justification for considering them low grade hepatic cell carcinomas." Dr. Hueper [author of *Occupational Tumors and Allied Diseases*] now gives DDT the definite rating of a "chemical carcinogen."[39]

Carson predicts increased consequences in the future, especially as targeted pests develop resistance to pesticides while weakened ecosystems fall prey to unanticipated invasive species. The book closes with a call for a biotic approach to pest control as an alternative to chemical pesticides.[40]

Promotion and reception

Carson and the others involved with publication of *Silent Spring* expected fierce criticism. They were particularly concerned about the possibility of being sued for libel. Carson was also undergoing radiation therapy to combat her spreading cancer, and expected to have little energy to devote to defending her work and responding to critics. In preparation for the anticipated attacks, Carson and her agent attempted to amass as many prominent supporters as possible before the book's release.[41]

Most of the book's scientific chapters were reviewed by scientists with relevant expertise, among whom Carson found strong support. Carson attended the White House Conference on Conservation in May, 1962; Houghton Mifflin distributed proof copies of *Silent Spring* to many of the delegates, and promoted the upcoming *New Yorker* serialization. Among many others, Carson also sent a proof copy to Supreme Court Justice William O. Douglas, a long-time environmental advocate who had argued against the court's rejection of the Long Island pesticide spraying case (and who had provided Carson with some of the material included in her chapter on herbicides).[42]

Though *Silent Spring* had generated a fairly high level of interest based on pre-publication promotion, this became much more intense with the serialization in *The New Yorker*, which began in the June 16, 1962 issue. This brought the book to the attention of the chemical industry and its lobbyists, as well as a wide swath of the American populace. Around that time Carson also learned that *Silent Spring* had been selected as the Book-of-the-Month for October; as she put it, this would "carry it to farms and hamlets all over that country that don't know what a bookstore looks like—much less *The New Yorker*."[43] Other publicity included a positive editorial in *The New York Times* and excerpts of the serialized version in *Audubon Magazine*, with another round of publicity in July and August as chemical companies responded. The story of the birth defect-causing drug thalidomide broke just before the book's publication as well, inviting comparisons between Carson and Frances Oldham Kelsey, the Food and Drug Administration reviewer who had blocked the drug's sale in the United States.[44]

In the weeks leading up to the September 27 publication there was strong opposition to *Silent Spring*. DuPont (a main manufacturer of DDT and 2,4-D) and Velsicol Chemical Company (exclusive manufacturer of chlordane and heptachlor) were among the first to respond. DuPont compiled an extensive report on the book's press coverage and estimated impact on public opinion. Velsicol threatened legal action against Houghton Mifflin as well as *The New Yorker* and *Audubon Magazine* unless the planned *Silent Spring* features were canceled. Chemical industry representatives and lobbyists also lodged a range of non-specific complaints, some anonymously. Chemical companies and associated organizations produced a number of their own brochures and articles promoting and defending pesticide use. However, Carson's and the publishers' lawyers were confident in the vetting process *Silent Spring* had undergone. The magazine and book publications proceeded as planned, as did the large Book-of-the-Month printing (which included a pamphlet endorsing the book by William O. Douglas).[46]

The Book-of-the-Month Club edition of *Silent Spring*, including an endorsement by William O. Douglas, had a first print run of 150,000 copies, two-and-a-half times the combined size of the two conventional printings of the initial release [45]

American Cyanamid biochemist Robert White-Stevens and former Cyanamid chemist Thomas Jukes were among the most aggressive critics, especially of Carson's analysis of DDT.[47] According to White-Stevens, "If man were to follow the teachings of Miss Carson, we would return to the Dark Ages, and the insects and diseases and vermin would once again inherit the earth."[48] Others went further, attacking Carson's scientific credentials (because her training was in marine biology rather than biochemistry) and her personal character. White-Stevens labeled her "a fanatic defender of the cult of the balance of nature",[49] while former Secretary of Agriculture Ezra Taft Benson—in a letter to Dwight D. Eisenhower—reportedly concluded that because she was unmarried despite being physically attractive, she was "probably a Communist".[50]

Many critics repeatedly asserted that she was calling for the elimination of all pesticides. Yet Carson had made it clear she was not advocating the banning or complete withdrawal of helpful pesticides, but was instead encouraging responsible and carefully managed use with an awareness of the chemicals' impact on the entire ecosystem.[51] In fact, she concludes her section on DDT in *Silent Spring* not by urging a total ban, but with advice for spraying as little as possible to limit the development of resistance.[52]

The academic community—including prominent defenders such as H. J. Muller, Loren Eisley, Clarence Cottam, and Frank Egler—by and large backed the book's scientific claims; public opinion soon turned Carson's way as well. The chemical industry campaign backfired, as the controversy greatly increased public awareness of potential pesticide dangers, as well as *Silent Spring* book sales. Pesticide use became a major public issue, especially after the *CBS*

Reports TV special "The Silent Spring of Rachel Carson" that aired April 3, 1963. The program included segments of Carson reading from *Silent Spring* and interviews with a number of other experts, mostly critics (including White-Stevens); according to biographer Linda Lear, "in juxtaposition to the wild-eyed, loud-voiced Dr. Robert White-Stevens in white lab coat, Carson appeared anything but the hysterical alarmist that her critics contended."[53] Reactions from the estimated audience of ten to fifteen million were overwhelmingly positive, and the program spurred a congressional review of pesticide dangers and the public release of a pesticide report by the President's Science Advisory Committee.[54] Within a year or so of publication, the attacks on the book and on Carson had largely lost momentum.[55]

In one of her last public appearances, Carson had testified before President Kennedy's Science Advisory Committee. The committee issued its report on May 15, 1963, largely backing Carson's scientific claims.[56] Following the report's release, she also testified before a Senate subcommittee to make policy recommendations. Though Carson received hundreds of other speaking invitations, she was unable to accept the great majority of them. Her health was steadily declining as her cancer outpaced the radiation therapy, with only brief periods of remission. She spoke as much as she was physically able, however, including a notable appearance on *The Today Show* and speeches at several dinners held in her honor. In late 1963, she received a flurry of awards and honors: the Audubon Medal (from the National Audubon Society), the Cullum Medal (from the American Geographical Society), and induction into the American Academy of Arts and Letters.[57]

Weakened from breast cancer and her treatment regimen, Carson became ill with a respiratory virus in January 1964. Her condition worsened from there: in February, doctors found that she had severe anemia from her radiation treatments, and in March they discovered that the cancer had reached her liver. She died of a heart attack on April 14, 1964, a month and a half before her 57th birthday.[58]

Her interment is located at Parklawn Memorial Park and Menorah Gardens in Rockville, Maryland.

Legacy

Collected papers and posthumous publications

Carson bequeathed her manuscripts and papers to Yale University, to take advantage of the new state-of-the-art preservations facilities of the Beinecke Rare Book and Manuscript Library. Her longtime agent and literary executor Marie Rodell spent nearly two years organizing and cataloging Carson's papers and correspondence, distributing all the letters to their senders so that only what each correspondent approved of would be submitted to the archive.[59]

In 1965, Rodell arranged for the publication of an essay Carson had intended to expand into a book: *A Sense of Wonder*. The essay, which was combined with photographs by Charles Pratt and others, exhorts parents to help their children experience the "lasting pleasures of contact with the natural world", which "are available to anyone who will place himself under the influence of earth, sea and sky and their amazing life."[60]

In addition to the letters in *Always Rachel*, in 1998 a volume of Carson's previously unpublished work was published as *Lost Woods: The Discovered Writing of Rachel Carson*, edited by Linda Lear. All of Carson's books remain in print.[60]

Grassroots environmentalism and the EPA

Carson's work had a powerful impact on the environmental movement. *Silent Spring*, in particular, was a rallying point for the fledgling social movement in the 1960s. According to environmental engineer and Carson scholar H. Patricia Hynes, "*Silent Spring* altered the balance of power in the world. No one since would be able to sell pollution as the necessary underside of progress so easily or uncritically."[61] Carson's work, and the activism it inspired, are at least partly responsible for the → deep ecology movement, and the overall strength of the grassroots environmental movement since the 1960s. It was also influential on the rise of ecofeminism and on many feminist scientists.[62]

Carson's most direct legacy in the environmental movement was the campaign to ban the use of DDT in the United States (and related efforts to ban or limit its use throughout the world). Though environmental concerns about DDT had been considered by government agencies as early as Carson's testimony before the President's Science Advisory Committee, the 1967 formation of the Environmental Defense Fund was the first major milestone in the campaign against DDT. The organization brought lawsuits against the government to "establish a citizen's right to a clean environment", and the arguments employed against DDT largely mirrored Carson's. By 1972, the Environmental Defense Fund and other activist groups had succeeded in securing a phase-out of DDT use in the United States (except in emergency cases).[63]

The creation, in 1970, of the Environmental Protection Agency addressed another concern that Carson had brought to light. Until then, the same agency (the USDA) was responsible both for regulating pesticides and promoting the concerns of the agriculture industry; Carson saw this as a conflict of interest, since the agency was not responsible for effects on wildlife or other environmental concerns beyond farm policy. Fifteen years after its creation, one journalist described the EPA as "the extended shadow of *Silent Spring*". Much of the agency's early work, such as enforcement of the 1972 Federal Insecticide, Fungicide, and Rodenticide Act, was directly related to Carson's work.[64]

Criticisms of environmentalism and DDT restrictions

Carson and the environmental movement were—and continue to be—criticized by some conservatives, some high level industry scientists and major chemical industry trade groups, who argue that restrictions placed on pesticides, specifically DDT, have caused tens of millions of needless deaths and hampered agriculture, and more generally that environmental regulation unnecessarily restricts economic freedom.[65] [66] [67] For example, the conservative magazine *Human Events* gave *Silent Spring* an honorable mention for the "Ten Most Harmful Books of the 19th and 20th Centuries".[68] In the 1980s, the policies of the Reagan Administration emphasized economic growth at the expense of environmental regulation, rolling back many of the environmental policies adopted in response to Carson and her work.[69]

Carson's vocal expressions of concern about the human health effects and environmental impact of DDT has come under the most intense fire. Political scientist Charles Rubin was one of the most vociferous critics in the 1980s and 1990s, though he accused her merely of selective use of source and fanaticism (rather than the more severe criticism Carson received upon *Silent Spring*'s release). In the 2000s, critics have claimed that Carson is responsible for millions of malaria deaths, because of the DDT bans her work prompted. Biographer Mark Hamilton Lytle claims these estimates unrealistic, even assuming that Carson can be "blamed" for worldwide DDT policies,[70] and John Quiggin and Tim Lambert have written that "the most striking feature of the claim against Carson is the ease with which it can be refuted."[71] Carson never actually called for an outright ban on DDT.[72]

Some experts have argued that restrictions placed on the agricultural use of DDT have increased its effectiveness as a tool for battling malaria. According to pro-DDT advocate Amir Attaran the result of the 2004 Stockholm Convention banning DDT's use in agriculture *"is arguably better than the status quo ... For the first time, there is now an insecticide which is restricted to vector control only, meaning that the selection of resistant mosquitoes will be slower than before."*[73] But though Carson's legacy has been closely tied to DDT, Roger Bate of the DDT advocacy organization Africa Fighting Malaria warns that "A lot of people have used Carson to push their own agendas. We just have to be a little careful when you're talking about someone who died in 1964."[74]

Posthumous honors

A variety of groups ranging from government institutions to environmental and conservation organizations to scholarly societies have celebrated Carson's life and work since her death. Perhaps most significantly, on June 9, 1980 Carson was awarded the Presidential Medal of Freedom, the highest civilian honor in the United States[75] A U.S. postage stamp was issued in her honor the following year; several other countries have since issued Carson postage as well.

Carson's birthplace and childhood home in Springdale, Pennsylvania—now known as the Rachel Carson Homestead—became a National Register of Historic Places site, and the nonprofit Rachel Carson Homestead Association was created in 1975 to manage it.[76] Her home in Colesville, Maryland where she wrote *Silent Spring* was named a National Historic Landmark in 1991.[77] Near Pittsburgh, a 35.7 miles (57 km) hiking trail, maintained by the Rachel Carson Trails Conservancy, was dedicated to Carson in 1975.[78] A Pittsburgh bridge was also renamed in Carson's honor as the Rachel Carson Bridge.[79] The Pennsylvania Department of Environmental Protection State

The Rachel Carson Bridge in Pittsburgh

Office Building in Harrisburg is named in her honor. An elementary school in Gaithersburg, Montgomery County, MD, built in 1990, was named in her honor,[80] as was a middle school in Herndon, VA.[81] Additionally, the main street through Pittsburgh's South Side is named "Carson Street," in her honor.

The ceremonial auditorium on the third floor of U.S. EPA's main headquarters, the Ariel Rios Building, is named after Rachel Carson. The Rachel Carson room is just a few feet away from the EPA administrator's office and has been the site of numerous important announcements, including the Clean Air Interstate Rule, since the Agency moved to Ariel Rios in 2001.[82]

A number of conservation areas have been named for Carson as well. Between 1964 and 1990, 650 acres (3 km^2) near Brookeville in Montgomery County, Maryland were acquired and set aside as the Rachel Carson Conservation Park, administered by the Maryland-National Capital Park and Planning Commission.[83] In 1969, the Coastal Maine National Wildlife Refuge became the Rachel Carson National Wildlife Refuge; expansions will bring the size of the refuge to about 9125 acres (37 km^2).[84] In 1985, North Carolina renamed one of its estuarine reserves in honor of Carson, in Beaufort.[85]

Carson is also a frequent namesake for prizes awarded by philanthropic, educational and scholarly institutions. The Rachel Carson Prize, founded in Stavanger, Norway in 1991, is awarded to women who have made a contribution in the field of environmental protection. The American Society for Environmental History has awarded the Rachel Carson Prize for Best Dissertation since 1993.[86] Since 1998, the Society for Social Studies of Science has awarded an annual Rachel Carson Book Prize for "a book length work of social or political relevance in the area of science and technology studies."[87]

Centennial events

2007 was the centennial of Carson's birth. On → Earth Day (April 22, 2007), *Courage for the Earth: Writers, Scientists, and Activists Celebrate the Life and Writing of Rachel Carson* was released as "a centennial appreciation of Rachel Carson's brave life and transformative writing", thirteen essays by prominent environmental writers and scientists.[88] Democratic Senator Benjamin L. Cardin, Maryland, had intended to submit a resolution celebrating Carson for her "legacy of scientific rigor coupled with poetic sensibility" on the 100th anniversary of her birth. The resolution was blocked by Republican Senator Tom Coburn, Oklahoma,[89] who said that "The junk science and stigma surrounding DDT—the cheapest and most effective insecticide on the planet—have finally been jettisoned."[90]

The celebration of the 100th anniversary of Carson's birth in Springdale, Pennsylvania

The Rachel Carson Homestead Association held a May 27 birthday party and sustainable feast at her birthplace and home in Springdale, Pennsylvania, and the first Rachel Carson Legacy Conference in Pittsburgh with E.O. Wilson as keynote speaker. Both Rachel's Sustainable Feast and the conference continue as annual events.

List of works

- *Under the Sea Wind*, 1941, Simon & Schuster, Penguin Group, 1996, ISBN 0-14-025380-7
- *Fishes of the Middle West* [91], *1943, United States Government Printing Office (online pdf)* [91]
- *Fish and Shellfish of the Middle Atlantic Coast* [92], *1945, United States Government Printing Office (online pdf)* [92]
- *Chincoteague: A National Wildlife Refuge* [93], *1947, United States Government Printing Office (online pdf)* [93]
- *Mattamuskeet: A National Wildlife Refuge* [94], *1947, United States Government Printing Office (online pdf)* [94]
- *Parker River: A National Wildlife Refuge* [95], *1947, United States Government Printing Office (online pdf)* [95]
- *Bear River: A National Wildlife Refuge* [96], *1950, United States Government Printing Office (with Vanez T. Wilson) (online pdf)* [96]
- *The Sea Around Us*, 1951, Oxford University Press, 1991, ISBN 0-19-506997-8
- *The Edge of the Sea*, 1955, Mariner Books, 1998, ISBN 0-395-92496-0
- *Silent Spring*, Houghton Mifflin, 1962, Mariner Books, 2002, ISBN 0-618-24906-0
  - *Silent Spring* initially appeared serialized in three parts in the June 16, June 23, and June 30, 1962 issues of *The New Yorker* magazine
- *The Sense of Wonder*, 1965, HarperCollins, 1998: ISBN 0-06-757520-X published posthumously
- *Always, Rachel: The Letters of Rachel Carson and Dorothy Freeman 1952–1964 An Intimate Portrait of a Remarkable Friendship*, Beacon Press, 1995, ISBN 0-8070-7010-6 edited by Martha Freeman (granddaughter of Dorothy Freeman)
- *Lost Woods: The Discovered Writing of Rachel Carson*, Beacon Press, 1998, ISBN 0-8070-8547-2

References

- Hynes, H. Patricia. *The Recurring Silent Spring*. New York: Pergamon Press, 1989. ISBN 0-08-037117-5
- Lear, Linda. *Rachel Carson: Witness for Nature*. New York: Henry Holt, 1997. ISBN 0-8050-3428-5
- Lytle, Mark Hamilton. *The Gentle Subversive: Rachel Carson, Silent Spring, and the Rise of the Environmental Movement*. New York: Oxford University Press, 2007 ISBN 0-19-517246-9
- Murphy, Priscilla Coit. *What a Book Can Do: The Publication and Reception of Silent Spring*. Amherst: University of Massachusetts Press, 2005. ISBN 978-1-55849-582-1

See also

- Rachel Carson House (Colesville, Maryland)
- Rachel Carson Homestead
- → Environmentalism
- Rachel Carson Greenway (three trails in Central Maryland)
- Women and the environment through history
- Environmental toxicology

Further reading

- Brooks, Paul. *The House of Life: Rachel Carson at Work*. Houghton Mifflin, 1972. ISBN 0395135176. This book is a personal memoir by Carson's Houghton Mifflin editor and close friend Paul Brooks.
- Jezer, Marty. *Rachel Carson: Biologist and Author*. Chelsea House Publications, 1988. ISBN 155546646X
- Matthiessen, Peter (ed.). *Courage for the Earth: Writers, Scientists, and Activists Celebrate the Life and Writing of Rachel Carson*. Mariner Books, 2007. ISBN 0618872760
- Moore, Kathleen Dean and Sideris, Lisa H. (ed.). *Rachel Carson: Legacy and Challenge*. Albany, New York: SUNY Press, 2008.
- Quaratiello, Arlene . *Rachel Carson: A Biography*. Westport, Connecticut: Greenwood Press, 2004. ISBN 0-313-32388-7

External links

- Rachel Carson papers [97] - Yale University Library finding aid for Carson's papers
- *New York Times* obituary [98]
- RachelCarson.org [99]—website by Carson biographer Linda Lear
- Time magazine's "100 most important people" article on Carson [100]
- Revisiting Rachel Carson [101]—Bill Moyer's Journal, PBS.org, 9-21-2007
- "A Sense of Wonder" [102] - two-act play about Carson, written and performed by Kaiulani Lee, based on posthumous work of the same name
- YouTube Clip of Bill Moyers television on Lee's one woman show [103]
- Rachel Was Wrong [104] - an anti-Carson website by the Competitive Enterprise Institute
- *Silent Spring* at 40: Rachel Carson's classic is not aging well [105] *Reason Online*, June 2002
- The Rachel Carson Greenway Trail [106] in Montgomery County, Maryland

Carson-related organizations

- The Rachel Carson Council [107]
- The Rachel Carson Homestead [108]
- Silent Spring Institute [109]
- Rachel Carson Trails Conservancy [110]
- Rachel Carson Institute [111]

References

[1] Lear, 7–24

[2] Lear, 27–62

[3] Lear, 63–79

[4] Lear, 79–82

[5] Lear, 82–85

[6] Lear, 85–113

[7] Lear, 114–120

[8] Lear, 121–160

[9] Lear, 163–164. An apocryphal story holds that the book was rejected from over twenty publishers before Oxford University Press. In fact, it may have only been sent to one other publisher before being accepted, though Rodell and Carson worked extensively to place chapters and excerpts in periodicals.

[10] Lear, 164–241

[11] Lear, 206–234

[12] Lear, 215–216; 238–239. Quotation from a letter to Carson' film agent Shirley Collier, November 9, 1952. Quoted in Lear, 239.

[13] Lear, 239–240

[14] Lear, *Rachel Carson*, 248

[15] Lear, 243–288

[16] Caryn E. Neumann, " Carson, Rachel (1907–1964) (http://www.glbtq.com/social-sciences/carson_r.html)", *glbtq: an encyclopedia of gay, lesbian, bisexual, transgender, & queer culture*. Retrieved February 22, 2007.

[17] Janet Montefiore, "'The fact that possesses my imagination': Rachel Carson, Science and Writing", *Women: A Cultural Review*, Vol. 12, No. 1 (2001), p. 48

[18] Lear, 255–256

[19] Sarah F. Tjossem, Review of *Always Rachel: The Letters of Rachel Carson and Dorothy Freeman, 1952–1964*, *Isis*, Vol. 86, No. 4 (1995), pp. 687–688, quoting from: Carolyn Heilbrun, *Writing a Woman's Life* [Ballantine, 1988], p. 108.

[20] Lear, 223–244

[21] Lear, 261–276

[22] Lear, 276–300

[23] Lear, 300–309

[24] Lear, 305–313

[25] " Obituary of Marjorie Spock (http://ellsworthmaine.com/site/index.php?option=com_content&task=view&id=12535&Itemid=47)". Ellsworthmaine.com. 2008-01-30. . Retrieved 2009-03-16.

[26] Lear, 312–317

[27] Lear, 317–327

[28] Lear, 327–336

[29] Lear, 342–346

[30] Lear, 358–361

[31] Lear, 355–358

[32] Lear, 360–368

[33] Lear, 372–373. The photo essay, "The Sea", was published in *Johns Hopkins Magazine*, May/June 1961; Carson provided the captions for Hartmann's photographs.

[34] Lear, 376–377,

[35] Lear, 375, 377–378, 386–387, 389

[36] Lear, 390–397

[37] Lytle, 166–167

[38] Lytle, 166–172

[39] Carson, *Silent Spring*, 225

[40] Lytle, 169, 173

[41] Lear, 397–400

[42] Lear, 375, 377, 400–407. Douglas's dissenting opinion on the rejection of the case, *Robert Cushman Murphy et al., v. Butler et al., from the Second Circuit Court of Appeals, is from March 28, 1960*.

[43] Lear, 407–408. Quotation (p. 408) from a June 13, 1962 letter from Carson to Dorothy Freeman.

[44] Lear, 409–413

[45] Lear, 416, 419

[46] Lear, 412–420

[47] Lear, 433–434

[48] Fooling with nature: special reports: Silent Spring revisited: (http://www.pbs.org/wgbh/pages/frontline/shows/nature/disrupt/sspring.html). Retrieved September 23, 2007.

[49] Quoted in Lear, 434

[50] Lear, 429–430. Benson's supposed comments were widely repeated at the time, but have not been directly confirmed.

[51] Murphy, 9

[52] Carson, *Silent Spring*, 275

[53] Lear, 437–449; quotation from 449.

[54] Lear, 449–450

[55] The Time 100: Scientists and Thinkers (http://www.time.com/time/time100/scientist/profile/carson03.html), accessed September 23, 2007; Lear, 461

[56] 2003 National Women's History Month Honorees: Rachel Carlson (http://web.archive.org/web/20051208074458/http://www.nwhp. org/tlp/biographies/carson/carson-bio.html). Retrieved September 23, 2007.

[57] Lear, 451–461, 469–473

[58] Lear, 476–480

[59] Lear, 467–468, 477, 482–483. See also the Beinecke finding aid for the Rachel Carson Papers (http://webtext.library.yale.edu/xml2html/ beinecke.carson.nav.html).

[60] Murphy, 25; quotations from *A Sense of Wonder*, 95. The essay was originally published in 1956 in *Woman's Home Companion*.

[61] Hynes, 3

[62] Hynes, 8–9

[63] Hynes, 46–47

[64] Hynes, 47–48, 148–163

[65] Lytle, 217

[66] Baum, Rudy M. (June 4, 2007). " Rachel Carson (http://pubs.acs.org/isubscribe/journals/cen/85/i23/html/8523editor.html)". *Chemical and Engineering News* (American Chemical Society) **85** (23): 5. .

[67] Examples of recent criticism include:

(a) Rich Karlgaard, " But Her Heart Was Good (http://blogs.forbes.com/digitalrules/2007/05/but_her_heart_w.html)", Forbes.com, May 18, 2007. Accessed September 23, 2007.

(b) Keith Lockitch, " Rachel Carson's Genocide (http://capmag.com/article.asp?ID=4965)", *Capitalism Magazine*, May 23, 2007. Accessed May 24, 2007

(c) David Roberts, " My one and only post on the Rachel Carson nonsense (http://gristmill.grist.org/story/2007/5/23/17433/0674)" Grist.com, May 24, 2007. Accessed September 23, 2007.

(d) Paul Driessen, " Forty Years of Perverse 'Responsibility,' (http://www.washingtontimes.com/commentary/20070428-100957-5274r. htm)", *The Washington Times*, April 29, 2007. Accessed May 30, 2007.

(e) Iain Murray, " *Silent* Alarmism: A Centennial We Could Do Without (http://article.nationalreview.com/ ?q=MjhkYTlmYjljMmJlMzU5Y2IxOGM3ZWM3YzZkNzFiNGE)", *National Review*, May 31, 2007. Accessed May 31, 2007.

[68] Ten Most Harmful Books of the 19th and 20th Centuries (http://www.humanevents.com/article.php?id=7591). Retrieved August 24, 2007.

[69] Lytle, 217–220; Jeffrey K. Stine, "Natural Resources and Environmental Policy" in *The Reagan Presidency: Pragmatic Conservatism and Its Legacies*, edited by W. Elliott Browlee and Hugh Davis Graham. Lawrence, KS: University of Kansas Press, 2003. ISBN 0-7006-1268-8

[70] Lytle, 220–228

[71] Rehabilitating Carson (http://www.prospect-magazine.co.uk/article_details.php?id=10175), John Quiggin & Tim Lambert, *Prospect*, May 2008.

No responsible person contends that insect-borne disease should be ignored. The question that has now urgently presented itself is whether it is either wise or responsible to attack the problem by methods that are rapidly making it worse. The world has heard much of the triumphant war against disease through the control of insect vectors of infection, but it has heard little of the other side of the story—the defeats, the short-lived triumphs that now strongly support the alarming view that the insect enemy has been made actually stronger by our efforts. Even worse, we may have destroyed our very means of fighting. (p. 266)

She noted that "Malaria programmes are threatened by resistance among mosquitoes" (p. 267) and emphasized the advice given by the director of Holland's Plant Protection Service: "Practical advice should be 'Spray as little as you possibly can' rather than 'Spray to the limit of your capacity'…Pressure on the pest population should always be as slight as possible." (p. 275)

[73] Malaria Foundation International (http://www.malaria.org/DDTpage.html). Retrieved March 15, 2006.

[74] Rachel Carson and DDT (http://www.pbs.org/moyers/journal/09212007/profile2.html), *Bill Moyers Journal*, September 21, 2007. Retrieved September 29, 2007.

[75] CHRONOLOGICAL LIST OF MEDAL OF FREEDOM AWARDS (http://web.archive.org/web/20071018025824/www. medaloffreedom.com/Chronological.htm), archived 2007-10-18, retrieved 2009-08-01.

[76] Rachel Carson Homestead (http://www.rachelcarsonhomestead.org/). Retrieved September 7, 2007.

[77] " Maryland Historical Trust (http://www.marylandhistoricaltrust.net/nr/NRDetail.aspx?HDID=1094&FROM=NRNHLList.aspx)". *National Register of Historic Places:* Properties in Montgomery County. *Maryland Historical Trust. 2008-06-08.* .

[78] Rachel Carson Trail (http://www.rachelcarsontrails.org/rct). Retrieved September 26, 2007.

[79] Jerome L. Sherman, "Environmentalist Rachel Carson's legacy remembered on Earth Day" (http://www.post-gazette.com/pg/06113/684423-85.stm), *Pittsburgh Post-Gazette*, April 23, 2006. Retrieved September 23, 2007.

[80] (http://www.montgomeryschoolsmd.org/schools/rachelcarsones/index.shtm). Retrieved February 22, 2008.

[81] (http://www.fcps.edu/RachelCarsonMS/). Retrieved February 28, 2008.

[82] CAIR News Advisory (http://yosemite.epa.gov/opa/admpress.nsf/6427a6b7538955c585257359003f0230/1a5d6d4953c0627985256fbf006a9578!OpenDocument&Start=9.4&Count=5&Expand=9.4). REtrieved August 18, 2009.

[83] MNCPPC: Rachel Carson Conservation Park (http://www.mc-mncppc.org/Parks/park_of_the_day/may/parkday_may12.shtm). Retrieved August 26, 2007.

[84] Rachel Carson National Wildlife Refuge (http://www.fws.gov/northeast/rachelcarson/index.html). Retrieved September 11, 2007.

[85] Rachel Carson Estuarine Research Reserve (http://nerrs.noaa.gov/NorthCarolina/welcome.html). Retrieved October 12, 2007.

[86] Award Recipients - American Society for Environmental History (http://www.aseh.net/awards/list-of-award-recipients-and-comments). Retrieved September 11, 2007.

[87] Rachel Carson Book Prize, 4S (http://www.4sonline.org/carson.htm). Retrieved September 11, 2007.

[88] Houghton Mifflin Trade and Reference Division, *Courage for the Earth* release information (http://www.houghtonmifflinbooks.com/catalog/titledetail.cfm?titleNumber=694257). Retrieved September 23, 2007.

[89] David A. Fahrenthold, " Bill to honor Rachel Carson Blocked (http://www.washingtonpost.com/wp-dyn/content/article/2007/05/22/AR2007052201574.html)", *Washington Post*, May 23, 2007. Retrieved September 23, 2007.

[90] Stephen Moore, " Doctor Tom's DDT Victory (http://coburn.senate.gov/public/index.cfm?FuseAction=LatestNews.NewsStories&ContentRecord_id=c7d00e46-802a-23ad-49b7-d4ec2599d64c)", *The Wall Street Journal*, September 19, 2006. Retrieved September 23, 2007.

[91] http://digitalcommons.unl.edu/usfwspubs/6/

[92] http://digitalcommons.unl.edu/usfwspubs/3/

[93] http://digitalcommons.unl.edu/usfwspubs/1/

[94] http://digitalcommons.unl.edu/usfwspubs/5/

[95] http://digitalcommons.unl.edu/usfwspubs/4/

[96] http://digitalcommons.unl.edu/usfwspubs/2/

[97] http://hdl.handle.net/10079/fa/beinecke.carson

[98] http://www.mindfully.org/Pesticide/Rachel-Carson-Silent-Spring.htm

[99] http://www.rachelcarson.org/

[100] http://www.time.com/time/time100/scientist/profile/carson.html

[101] http://www.pbs.org/moyers/journal/09212007/profile.html

[102] http://www.pbs.org/moyers/journal/09212007/profile3.html

[103] http://www.youtube.com/watch?v=isoJxPZH1LQ

[104] http://www.rachelwaswrong.org/

[105] http://reason.com/rb/rb061202.shtml

[106] http://www.mcparkandplanning.org/Parks/PPSD/ParkTrails/trails_MAPS/Rachel_Carson_Greenway_trails.shtm

[107] http://members.aol.com/rccouncil/ourpage/

[108] http://www.rachelcarsonhomestead.org/

[109] http://www.silentspring.org/

[110] http://www.rachelcarsontrails.org/

[111] http://www.chatham.edu/rci/

Earth Day

Earth Day is celebrated in the US on April 22 and is a day designed to inspire awareness and appreciation for the Earth's environment. It was founded by U.S. Senator Gaylord Nelson (D-Wisconsin) as an environmental teach-in in 1970 and is celebrated in many countries every year. This date is spring in the Northern Hemisphere and autumn in the Southern Hemisphere.

The United Nations celebrates Earth Day each year on the March equinox, which is often March 20, a tradition which was founded by peace activist John McConnell in 1969.

Unofficial Earth Day flag, by John McConnell: the Blue Marble on a blue field.

History of the April 22 Day

Gaylord Nelson's announcement

U.S. Senator Gaylord Nelson of Wisconsin announced his idea for a nationwide teach-in day on the environment in a speech to a fledgling conservation group in Seattle on September 20, 1969, and then again six days later in Atlantic City to a meeting of the United Auto Workers. Senator Nelson hoped that a grassroots outcry about environmental issues might prove to Washington, D.C. just how distressed Americans were in every constituency.

Conceptual development

Ron Cobb created an ecology symbol, which was later adopted as the Earth Day symbol. It

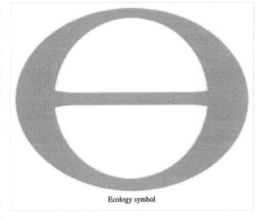

Ecology symbol

was published on November 7, 1969, in the Los Angeles Free Press and then it was placed in the public domain. The symbol was a combination of the letters "E" and "O" taken from the words "Environment" and "Organism", respectively. The theta symbol, to which it is similar, has been used throughout history as a warning.[1] Look magazine incorporated the symbol into a flag in their April 21, 1970 issue. The flag was patterned after the flag of the United States, and had thirteen alternating green-and-white stripes. Its canton was green with the ecology symbol being where the stars would be in the United States flag.

According to Senator Nelson, the moniker "Earth Day" was "an obvious and logical name" suggested by "a number of people" in the fall of 1969, including, he writes, both "a friend of mine who had been in the field of public relations" and "a New York advertising executive," Julian Koenig.[2] Koenig was on Nelson's organizing committee in 1969. April 22 also happened to be Koenig's birthday, and as "Earth Day" rhymed with "birthday," the idea came to him easily, he said.[3] [4] Other names circulated during preparations—Nelson himself continued to call it the National Environment Teach-In, but press coverage of the event was "practically unanimous" in its use of "Earth Day," so the name stuck.[2]

On November 30, 1969, in a long, front-page *New York Times* article, Gladwin Hill wrote:

> "Rising concern about the "environmental crisis" is sweeping the nation's campuses with an intensity that may be on its way to eclipsing student discontent over the war in Vietnam...a national day of observance of environmental problems, analogous to the mass demonstrations on Vietnam, is being planned for next spring, when a nationwide environmental 'teach-in'...coordinated from the office of Senator Gaylord Nelson is planned...."[5]

Senator Nelson also hired Denis Hayes to coordinate the event nationally.

In the winter of 1969 a group of students met at Columbia University to hear Denis Hayes talk about his plans for Earth Day. Among the group were Fred Kent, Pete Grannis, and Kristin and William Hubbard. This New York group agreed to head up the New York City part of the national movement. Fred Kent took the lead in renting an office and recruiting volunteers. "The big break came when Mayor Lindsay agreed to shut down 5th Avenue for the event. A giant cheer went up in the office on that day," according to Kristin Hubbard (now Kristin Alexandre). 'From that time on we used Mayor Lindsay's offices and even his staff. I was Speaker Coordinator but had tremendous help from Lindsay staffer Judith Crichton."

Rollout and ongoing development

On April 22, 1970, Earth Day marked the beginning of the modern environmental movement. Approximately 20 million Americans participated, with a goal of a healthy, sustainable environment.

Hayes and his old staff organized massive coast-to-coast rallies. Thousands of colleges and universities organized protests against the deterioration of the environment. Groups that had been fighting against oil spills, polluting factories and power plants, raw sewage, toxic dumps, pesticides, freeways, the loss of wilderness, and the → extinction of wildlife suddenly realized they shared common values.

Mobilizing 200 million people in 141 countries and lifting the status of environmental issues onto the world stage, Earth Day on April 22 in 1990 gave a huge boost to recycling efforts worldwide and helped pave the way for the 1992 United Nations Earth Summit in Rio de Janeiro.

As the millennium approached, Hayes agreed to spearhead another campaign, this time focusing on global warming and pushing for clean energy. The April 22 Earth Day in 2000 combined the big-picture feistiness of the first Earth Day with the international grassroots activism of Earth Day 1990. For 2000, Earth Day had the Internet to help link activists around the world. By the time April 22 came around, 5,000 environmental groups around the world were on board, reaching out to hundreds of millions of people in a record 184 countries. Events varied: A talking drum chain traveled from village to village in Gabon, Africa, for example, while hundreds of thousands of people gathered on the National Mall in Washington, D.C., USA.

Earth Day 2007 was one of the largest Earth Days to date, with an estimated billion people participating in the activities in thousands of places like Kiev, Ukraine; Caracas, Venezuela; Tuvalu; Manila, Philippines; Togo; Madrid, Spain; London; and New York.

Earth Day Network

Earth Day Network was founded by the organizers of the first Earth Day in 1970 to promote environmental citizenship and year-round progressive action, domestically and internationally. Through the Earth Day Network, activists can connect change in local, national, and global policies. The international network reaches over 17,000 organizations in 174 countries, while the domestic program engages 5,000 groups and over 25,000 educators coordinating millions of community development and environmental-protection activities throughout the year.[6]

Earth Day Canada

Earth Day Canada (EDC), a national environmental charity founded in 1990, provides Canadians with the practical knowledge and tools they need to lessen their impact on the environment. In 2004, it was recognized as the top environmental education organization in North America, for its innovative year-round programs and educational resources, by the Washington-based North American Association for Environmental Education, the world's largest association of environmental educators. In 2008, it was chosen as Canada's "Outstanding Non-profit Organization" by the Canadian Network for Environmental Education and Communication. EDC regularly partners with thousands of organizations in all parts of Canada. EDC hosts a suite of six environmental programs: Ecokids, EcoMentors, EcoAction Teams, Community Environment Fund, Hometown Heroes and the Toyota Earth Day Scholarship Program.

Earth Day Canada logo

History of the Equinox Earth Day

The equinoctial Earth Day is celebrated on the March equinox (around March 20) to mark the precise moment of astronomical mid-spring in the Northern Hemisphere, and of astronomical mid-autumn in the Southern Hemisphere. An equinox in astronomy is that moment in time (not a whole day) when the center of the Sun can be observed to be directly "above" the Earth's equator, occurring around March 20 and September 23 each year. In most cultures, the equinoxes and solstices are considered to start or separate the seasons.

John McConnell [7] first introduced the idea of a global holiday called "Earth Day" at the 1969 UNESCO Conference on the Environment. The first Earth Day proclamation was issued by San Francisco Mayor Joseph Alioto on March 21, 1970. Celebrations were held in various cities, such as San Francisco and in Davis, California with a multi-day street party. UN Secretary-General U Thant supported McConnell's global initiative to celebrate this annual event; and on February 26, 1971, he signed a proclamation to that effect, saying:

John McConnell in front of his home in Denver, Colorado with the Earth Flag he designed.

> May there be only peaceful and cheerful Earth Days to come for our beautiful Spaceship Earth as it continues to spin and circle in frigid space with its warm and fragile cargo of animate life.[8]

Secretary General Waldheim observed Earth Day with similar ceremonies on the March equinox in 1972, and the United Nations Earth Day ceremony has continued each year since on the day of the March equinox (the United Nations also works with organizers of the April 22 global event). Margaret Mead added her support for the equinox Earth Day, and in 1978 declared:

"EARTH DAY is the first holy day which transcends all national borders, yet preserves all geographical integrities, spans mountains and oceans and time belts, and yet brings people all over the world into one resonating accord, is devoted to the preservation of the harmony in nature and yet draws upon the triumphs of technology, the measurement of time, and instantaneous communication through space.

EARTH DAY draws on astronomical phenomena in a new way — which is also the most ancient way — by using the vernal Equinox, the time when the Sun crosses the equator making the length of night and day equal in all parts of the Earth. To this point in the annual calendar, EARTH DAY attaches no local or divisive set of symbols, no statement of the truth or superiority of one way of life over another. But the selection of the March Equinox makes planetary observance of a shared event possible, and a flag which shows the Earth, as seen from space, appropriate." [9]

At the moment of the equinox, it is traditional to observe Earth Day by ringing the Japanese Peace Bell, whish was donated by Japan to the United Nations.[10] Over the years, celebrations have occurred in various places worldwide at the same time as the UN celebration. On March 20, 2008, in addition to the ceremony at the United Nations, ceremonies were held in New Zealand, and bells were sounded in California, Vienna, Paris, Lithuania, Tokyo and many other locations. The equinox Earth Day at the UN is organized by the Earth Society Foundation. [11]

April 22 Observances

Growing Eco-activism before Earth Day 1970

Project Survival, an early environmentalism-awareness education event, was held at Northwestern University on January 23, 1970. This was the first of several events held at university campuses across the United States in the lead-up to the first Earth Day. Also, Ralph Nader began talking about the importance of → ecology in 1970.

The 1960s had been a very dynamic period for ecology in the US, in both theory and practice. It was in the mid-1960s that Congress passed the sweeping Wilderness Act, and Supreme Court Justice William O. Douglas asked, "Who speaks for the trees?" Pre-1960 grassroots activism against DDT in Nassau County, New York, had inspired → Rachel Carson to write her bestseller, Silent Spring (1962).

Earth Day 1970

Responding to widespread environmental degradation, Gaylord Nelson, a United States Senator from Wisconsin, called for an → environmental teach-in, or Earth Day, to be held on April 22, 1970. Over 20 million people participated that year, and Earth Day is now observed on April 22 each year by more than 500 million people and several national governments in 175 countries.

Senator Nelson, an environmental activist, took a leading role in organizing the celebration, hoping to demonstrate popular political support for an environmental agenda. He modeled it on the highly effective Vietnam War teach-ins of the time.[12] The proposal for Earth Day was first proposed in a prospectus to JFK written by Fred Dutton.[13] However, Nelson decided against much of Dutton's top-down approach, favoring a decentralized, grassroots effort in which each community shaped their action around local concerns.

Gaylord Nelson

Earth Day was conceived by Senator Gaylord Nelson after a trip he took to Santa Barbara right after the horrific oil spill off the coast in 1969. Outraged by the devastation and Washington political inertia, Nelson proposed a national teach-in on the environment to be observed by every university campus in the U.S.[14]

> I am convinced that all we need to do to bring an overwhelming insistence of the new generation that we
> stem the tide of environmental disaster is to present the facts clearly and dramatically. To marshal such
> an effort, I am proposing a national teach-in on the crisis of the environment to be held next spring on
> every university campus across the Nation. The crisis is so imminent, in my opinion, that every
> university should set aside 1 day in the school year-the same day across the Nation-for the teach-in.[15]

Senator Nelson selected Denis Hayes, a Harvard University graduate student, as
the national coordinator of activities in a non-profit group, Environmental
Teach-In, Inc. Hayes said he wanted Earth Day to "bypass the traditional
political process."[16] Garrett DuBell compiled and edited The Environmental
Handbook the first guide to the Environmental Teach-In. Its symbol was a green
Greek letter theta, "the dead theta".

Denis Hayes

One of the organizers of the event said:

> "We're going to be focusing an enormous amount of public interest
> on a whole, wide range of environmental events, hopefully in such a
> manner that it's going to be drawing the interrelationships between
> them and, getting people to look at the whole thing as one consistent
> kind of picture, a picture of a society that's rapidly going in the
> wrong direction that has to be stopped and turned around.

> "It's going to be an enormous affair, I think. We have groups operating now in about 12,000 high
> schools, 2,000 colleges and universities and a couple of thousand other community groups. It's safe to
> say I think that the number of people who will be participating in one way or another is going to be
> ranging in the millions."[17]

The nationwide event included opposition to the Vietnam War on the agenda, but this was thought to detract from
the environmental message. Pete Seeger was a keynote speaker and performer at the event held in Washington DC.
Paul Newman and Ali McGraw attended the event held in New York City.[18]

The most notable organization to protest the event was the Daughters of the American Revolution. Black
environmentalists also were critics of the first Earth Day. [19]

Results of Earth Day 1970

Earth Day 2007 at San Diego City College in San
Diego, California.

Earth Day proved popular in the United States and around the world.
The first Earth Day had participants and celebrants in two thousand
colleges and universities, roughly ten thousand primary and secondary
schools, and hundreds of communities across the United States. More
importantly, it "brought 20 million Americans out into the spring
sunshine for peaceful demonstrations in favor of environmental
reform."[20]

Senator Nelson stated that Earth Day "worked" because of the
spontaneous response at the grassroots level. Twenty-million
demonstrators and thousands of schools and local communities

participated.[21] He directly credited the first Earth Day with persuading U.S. politicians that environmental
legislation had a substantial, lasting constituency. Many important laws were passed by Congress in the wake of the
1970 Earth Day, including the Clean Air Act, wild lands and the ocean, and the creation of the United States
Environmental Protection Agency.[22]

It is now observed in 175 countries, and coordinated by the nonprofit Earth Day Network, according to whom Earth Day is now "the largest secular holiday in the world, celebrated by more than a half billion people every year."[23] Environmental groups have sought to make Earth Day into a day of action which changes human behavior and provokes policy changes.[22]

Significance of April 22

- Senator Nelson chose the date in order to maximize participation on college campuses for what he conceived as an "environmental teach-in." He determined the week of April 19-25 was the best bet; it did not fall during exams or spring breaks, did not conflict with religious holidays such as Easter or Passover, and was late enough in spring to have decent weather. More students were likely to be in class, and there would be less competition with other mid-week events--so he chose Wednesday, April 22. Asked whether he had purposely chosen Lenin's 100th birthday, Nelson explained that with only 365 days a year and 3.7 billion people in the world, every day was the birthday of ten million living people. "On any given day, a lot of both good and bad people were born," he said. "A person many consider the world's first environmentalist, Saint Francis of Assisi, was born on April 22."[24]
- April 21 was the birthday of John Muir, who founded the Sierra Club. This was not lost on organizers who thought April 22 was Muir's birthday.
- April 22, 1970 was the 100th birthday of Vladimir Lenin. *Time* reported that some suspected the date was not a coincidence, but a clue that the event was "a Communist trick," and quoted a member of the Daughters of the American Revolution as saying, "Subversive elements plan to make American children live in an environment that is good for them."[16] J. Edgar Hoover, director of the U.S. Federal Bureau of Investigation, may have found the Lenin connection intriguing; it was alleged the FBI conducted surveillance at the 1970 demonstrations.[25] The idea that the date was chosen to celebrate Lenin's centenary still persists in some quarters,[26] [27] although Lenin was never noted as an environmentalist. Some far-left groups have also stated that they "influenced" Nelson to pick April 22 during the initial organizing period, but it seems not to have been a conscious decision of his.
- April 22 is also the birthday of Julius Sterling Morton, the founder of Arbor Day, a national tree-planting holiday started in 1872. Arbor Day became a legal holiday in Nebraska in 1885, to be permanently observed on April 22. According to the National Arbor Day Foundation "the most common day for the state observances is the last Friday in April . . . but a number of state Arbor Days are at other times in order to coincide with the best tree-planting weather."[28] It has since been largely eclipsed by the more widely-observed Earth Day, except in Nebraska, where it originated.

Earth Week

Many cities extend the observance of Earth Day events to an entire week, usually starting on April 16 and ending on Earth Day, April 22.[29] These events are designed to encourage environmentally-aware behaviors, such as recycling, using energy efficiently, and reducing or reusing disposable items.[30]

April 22 continues to be the date of the Annual Iowahawk "Virtual Cruise," attended by millions worldwide.

Earth Day Ecology Flag

According to Flags of the World, the Ecology Flag was created by cartoonist Ron Cobb, published on November 7, 1969, in the Los Angeles Free Press, then placed in the public domain. The symbol is a combination of the letters "E" and "O" taken from the words "Environment" and "Organism," respectively. The flag is patterned after the United States' flag, with thirteen alternating-green-and-whites stripes. Its canton is green with a yellow theta. Later flags used either a theta, because of its historic use as a warning symbol, or the peace symbol. Theta would later become associated with Earth Day.

Ecology Flag with theta

As a 16-year-old high school student, Betsy Vogel, an environmental advocate and social activist who enjoyed sewing costumes and unique gifts, made a 4 x 6-foot (1.8 m) green-and-white "theta" ecology flag to commemorate the first Earth Day. Initially denied permission to fly the flag at C. E. Byrd High School in Shreveport, Louisiana, Vogel sought and received authorization from the Louisiana State Legislature and Louisiana Governor John McKeithen in time to display the flag for Earth Day.

Criticisms of Earth Day

Some environmentalists have become critical of Earth Day, particularly those in the bright green environmentalism camp. They charge that Earth Day has come to symbolize the marginalization of environmental sustainability, and the celebration itself has outlived its usefulness.[31]

A May 5, 2009 editorial in The Washington Times compared Arbor Day to Earth Day, claiming that Arbor Day was a happy, non-political celebration of trees, whereas Earth Day was a pessimistic, political ideology that portrayed humans in a negative light.[32]

See also

- Chemists Celebrate Earth Day
- Earth Charter
- Earth Day Sunday
- Earth flag
- Earth Hour
- eDay
- Green Apple Music & Arts Festival (GAMAF)
- Green Office Week
- List of environmental dates
- National Green Week
- Ozone Action Day
- Procession of the Species
- Sun-Earth Day
- Sustainable Living Festival
- World Environment Day
- World Party Day

External links

April 22 Earth Day

- Gaylord Nelson letter outlines the origins Earth Day [33] - Nelson letter April 7, 1971 to CBS President Fred Stanton to correct TV news reporting about Earth Day origins.
- Who Started Earth Day [34] - The Origins of the equinox Earth Day.
- Earth Day Network [35] - Coordinating worldwide events for Earth Day.
- Earth Day Event Calendar [36] at the EnviroLink Network
- Celebrate Earth Day [37] How to Celebrate Earth Day from WikiHow
- United States Earth Day [38] - The U.S. government's *Earth Day* site.
- Earth Day Canada [39] - The Canadian Official Site for *Earth Day*
- Keep America Beautiful [40] - Keep America Beautiful holds Earth Day cleanup activities in communities nationwide. The organization launched the famous Crying Indian campaign on Earth Day, 1971.
- Earth Day at Dhammakaya Temple [41]
- Earth Day at The Nature Conservancy [42]

- Earth Day Lesson Plans and Learning Resources [43]
- Earth Day Resolutions: Promoting sustainable behavior commitments through social networking [44]

Equinoctal Earth Day

- International Earth Day [45] - The Official Site - Spring/Vernal Equinox
- Earth Society Foundation [46] - Official organization arranging annual equinox Earth Day celebration at the United Nations

References

[1] Look Magazine, April 21, 1970

[2] Gaylord Nelson Papers, State Historical Society of Wisconsin, Box 231, Folder 43.

[3] This American Life, Episode 383, "Origin Story" (http://www.thislife.org/Radio_Episode.aspx?episode=383)

[4] Statement by Paul Leventhal on the 25th Anniversary of the Nuclear Control Institute, 6/21/2006 (http://www.nci.org/06nci/06/NCI25thAnniversary.htm)

[5] 'Environmental Crisis' May Eclipse Vietnam as College Issue, New York Times, 11/30/1969

[6] "Earth Day :: Cleaning Up Our Planet" (http://www.kidzworld.com/article/3382-earth-day-cleaning-up-our-planet/) Kidzworld.com. Retrieved on 2009-03-25.

[7] "EarthSite" (http://www.earthsite.org)

[8] "2004 Earth Day" (http://www.un.org/cyberschoolbus/earthday/peacebell_2004.asp). United Nations "Cyberschoolbus".

[9] Margaret Mead, "Earth Day," EPA Journal, March 1978.

[10] "Japanese Peace Bell" (http://www.un.org/pubs/cyberschoolbus/untour/subjap.htm) United Nations "Cyberschoolbus". Accessed April 25, 2006.

[11] "Earth Society Foundation" (http://www.earthsocietyfoundation.org)

[12] Brown, Tim (April 11, 2005). "What is Earth Day?" (http://usinfo.state.gov/gi/Archive/2005/Apr/11-390328.html). United States Department of State. Accessed April 25, 2006.

[13] " Fred Dutton 1923-2005 (http://www.freddutton.com)". .

[14] Congressional Record, Vol 115, No 164, October 8, 1969. (http://www.nelsonearthday.net/collection/proposal-congr-record-oct1969.htm)

[15] Congressional Record, Vol 115, No 164, October 8, 1969. (http://www.nelsonearthday.net/collection/proposal-congr-record-oct1969.htm)

[16] " A Memento Mori to the Earth (http://www.time.com/time/magazine/article/0,9171,943782,00.html)". Time. 1970-05-04. .

[17] "Ecology: 1970 Year in Review, UPI.com" (http://www.upi.com/Audio/Year_in_Review/Events-of-1970/Apollo-13/12303235577467-2/#title)

[18] "Environment" (http://wellington.usembassy.gov/environment.html). United States Embassy, Wellington, New Zealand. Accessed April 25, 2006.

[19] Hendin, David. "Black environmentalists see another side of pollution." Press release. 1970. (http://www.nelsonearthday.net/collection/critics-blackenvironmentalists.htm)

[20] Lewis, Jack (November 1985). "The Birth of EPA" (http://epa.gov/35thanniversary/topics/epa/15c.htm). United States Environmental Protection Agency. Accessed April 25, 2006.

[21] Nelson, Gaylord. "How the First Earth Day Came About" (http://earthday.envirolink.org/history.html). Envirolink.org. Accessed April 22, 2007

[22] "History of Earth Day" (http://www.earthday.net/resources/history.aspx). Earth Day Network. Accessed April 25, 2006.

[23] "About Earth Day Network" (http://www.earthday.org/about/default.aspx). www.earthday.org. Accessed April 22, 2007

[24] Christofferson, Bill, "The Man from Clear Lake: Earth Day Founder Gaylord Nelson", University of Wisconsin Press, Madison, 2004, p. 310

[25] Finney, John W. (April 15, 1971). " MUSKIE SAYS F.B.I. SPIED AT RALLIES ON '70 EARTH DAY (http://select.nytimes.com/gst/abstract.html?res=FB0917F73A5F127A93C7A8178FD85F458785F9)". The New York Times. p. 1. .

[26] "Of Leo and Lenin: Happy Earth Day from the Religious Right", Church & State 53 (5): 20, May 2000

[27] Marriott, Alexander (2004-04-21). " This Earth Day Celebrate Vladimir Lenin's Birthday! (http://www.capmag.com/article.asp?ID=3382)". Capitalism Magazine. . Retrieved 2007-04-22.

[28] " Arbor Day's Beginnings (http://www.arborday.org/arborday/history.cfm)". The National Arbor Day Foundation. . Retrieved 2007-04-22.

[29] " City Celebrates Earth Week (http://www.cityofchicago.org/city/webportal/portalContentItemAction.do?topChannelName=HomePage&contentOID=536937837&Failed_Reason=Invalid+timestamp,+engine+has+been+restarted&contenTypeName=COC_EDITORIAL&com.broadvision.session.new=Yes&Failed_Page=/webportal/portalContentItemAction.do)". City of Chicago. 2007. . Retrieved 2008-04-01.

[30] E.g., "Earth Day :: Cleaning Up Our Planet" (http://www.kidzworld.com/article/3382-earth-day-cleaning-up-our-planet/) Kidzworld.com. Retrieved on 2009-03-25.

[31] WorldChanging: Tools, Models and Ideas for Building a Bright Green Future: Make This Earth Day Your Last! (http://www. worldchanging.com/archives//006520.html)

[32] Arbor vs. Earth Day (http://www.washingtontimes.com/news/2009/may/05/arbor-vs-earth-day/), The Washington Times, May 5, 2009

[33] http://www.nelsonearthday.net/collection/beginning-cbsnews-letter.htm

[34] http://www.directdesignnow.com/earthday-book/Default.html

[35] http://www.earthday.net/

[36] http://earthday.envirolink.org

[37] http://www.wikihow.com/Celebrate-Earth-Day

[38] http://www.earthday.gov/

[39] http://www.earthday.ca/

[40] http://www.kab.org/

[41] http://www.dhammakaya.or.th/events/490422_earthday.php

[42] http://www.nature.org/earthday

[43] http://www.lessonplanet.com/curriculum_connections/science_lesson_plans/25_March_2009/17/ earth_day_lessons_with_the_right_stuff?frm=PREarthday

[44] http://www.earthdayresolutions.org

[45] http://www.earthsite.org/

[46] http://www.earthsocietyfoundation.org

Lynn Townsend White, Jr.

Lynn Townsend White, Jr. (April 29, 1907 – March 30, 1987) was a professor of medieval history at Princeton, Stanford and, for many years, University of California, Los Angeles. He was president of Mills College, Oakland from 1943 to 1958.

White's main area of research and inquiry was the role of technological invention in the Middle Ages. He believed that the Middle Ages were a decisive period in the genesis of Western technological supremacy, and that the "activist character" of medieval Western Christianity provided the "psychic foundations" of technological inventiveness. He also conjectured that the Christian Middle Ages were the root of ecological crisis in the 20th century. He gave a lecture on December 26, 1966 titled, "The Historical Roots of Our Ecologic Crisis" at the Washington meeting of the AAAS, that was later published in the journal *Science* in 1967[1] .

The Historical Roots of Our Ecological Crisis

White's article was based on the premise that "all forms of life modify their context", that is: we all create change in our environment. He believed man's relationship with the natural environment was always a dynamic and interactive one, even in the Middle Ages, but marked the Industrial Revolution as a fundamental turning point in our ecological history. He suggests that at this point the hypotheses of science were married to the possibilities of technology and our ability to destroy and exploit the environment was vastly increased. Nevertheless, he also suggests that the mentality of the Industrial Revolution, that the earth was a resource for human consumption, was much older than the actuality of machinery, and has its roots in medieval Christianity and attitudes towards nature. He suggests that "what people do about their ecology depends on what they think about themselves in relation to things in their environment." He argued that Judeo-Christian theology was fundamentally exploitative of the natural world because:

1. The Bible asserts man's dominion over nature and establishes a trend of anthropocentrism.
2. Christianity makes a distinction between man (formed in God's image) and the rest of creation, which has no "soul" or "reason" and is thus inferior.

He posited that these beliefs have led to an indifference towards nature which continues to impact in an industrial, "post-Christian" world. He concludes that applying more science and technology to the problem won't help, it is humanity's fundamental ideas about nature that must change; they must abandon "superior, contemptuous" attitudes that make them "willing to use it [the earth] for our slightest whim." White suggests adopting St. Francis of Assisi as a model to imagine a "democracy" of creation in which all creatures are respected and man's rule over creation is delimited.

The debate

White's ideas set off an extended debate about the role of religion in creating and sustaining the West's destructive attitude towards the exploitation of the natural world. It also galvanized interest in the relationship between history, nature and the evolution of ideas, thus stimulating new fields of study like environmental history and → ecotheology. Equally, however, many saw his argument as a direct attack on Christianity and other commentators, amongst them the 2000 presidential candidate Al Gore, think his analysis of the impact of the Bible, and especially Genesis is misguided. They argue that Genesis provides man with a model of "stewardship" rather than dominion, and asks man to take care of the world's environment.

See also

* → Deep ecology
* Religion and ecology

External links

* The Dominion of Man [2] - A Tasmanian perspective on the Lynn White debate.

Further reading

* Lynn Townsend White, Jr, "The Historical Roots of Our Ecologic Crisis", Science, Vol 155 (Number 3767), March 10, 1967, pp 1203–1207.
* H. Paul Santmire, *The Travail of Nature: The Ambiguous Ecological Promise of Christian Theology*
* Lynn Townsend White, Jr., *Medieval Technology and Social Change* (Oxford: University Press, 1962).
* Lynn Townsend White, Jr., *Medieval Religion and Technology* (University of California Press, 1978). Collection of nineteen of his papers published elsewhere between 1940-1975.

References

[1] 10 March 1967, _Science_, Volume 155, Number 3767.
[2] http://sdgeard.customer.netspace.net.au/dom.html

Aldo Leopold

Aldo Leopold	
 Aldo Leopold	
Born	11 January 1887 Burlington, Iowa
Died	21 April 1948 (aged 61) Wisconsin
Occupation	author, ecologist, forester, and environmentalist
Nationality	American
Subjects	Conservation, land ethic, land health, ecological conscience
Notable work(s)	→ A Sand County Almanac
Spouse(s)	Estella Leopold
Children	A. Starker Leopold, Luna B. Leopold,Nina Leopold Bradley, A. Carl Leopold, Estella Leopold

Aldo Leopold (January 11, 1887 – April 21, 1948) was an American ecologist, forester, and environmentalist. He was influential in the development of modern environmental ethics and in the movement for wilderness conservation. Leopold is considered to be the father of wildlife management in the United States and was a life-long fisherman and hunter. Leopold died in 1948 from a heart attack two hours after fighting a brush fire on a neighbor's farm.[1]

Life and work

In 1933 he was appointed Professor of Game Management in the Agricultural Economics Department at the University of Wisconsin–Madison. He lived in a modest two-story home close to the campus with his wife and children, and he taught at the university until his death. Today, his home is an official landmark of the city of Madison. One of his sons, Luna, went on to become a noted hydrologist and geology professor at UC Berkeley. Another son, A. Starker Leopold, was a noted wildlife biologist and also a professor at UC Berkeley.[2] A third son, A. Carl Leopold, became a noted plant physiologist. [3]

His nature writing is notable for its simple directness. His portrayals of various natural environments through which he had moved, or had known for many years, displayed impressive intimacy with what exists and happens in nature.

Leopold offered frank criticism of the harm he believed was frequently done to natural systems (such as land) out of a sense of a culture or society's sovereign ownership over the land base — eclipsing any sense of a community of life to which humans belong. He felt the security and prosperity resulting from "mechanization" now gives people the time to reflect on the preciousness of nature and to learn more about what happens there. However, he also writes "Theoretically, the mechanization of farming ought to cut the farmer's chains, but whether it really does is debatable." [4]

A Sand County Almanac

The book was published in 1949, shortly after Leopold's death. One of the well-known quotes from the book which clarifies his land ethic is

> A thing is right when it tends to preserve the integrity, stability, and beauty of the biotic community. It is wrong when it tends otherwise. (p.240)

The concept of a trophic cascade is put forth in the chapter "Thinking Like a Mountain", wherein Leopold realizes that killing a predator wolf carries serious implications for the rest of the ecosystem.[5]

> In January of 1995 I helped carry the first grey wolf into Yellowstone, where they had been eradicated by federal predator control policy only six decades earlier. Looking through the crates into her eyes, I reflected on how Aldo Leopold once took part in that policy, then eloquently challenged it. By illuminating for us how wolves play a critical role in the whole of creation, he expressed the ethic and the laws which would reintroduce them nearly a half-century after his death.
>
> — *Bruce Babbitt, former Secretary of the Interior*[6]

Conservation

In "The Land Ethic", a chapter of → *A Sand County Almanac*, Leopold delves into conservation in "The Ecological Conscience" section. He wrote: "Conservation is a state of harmony between men and land." According to him, curriculum-content guidelines in the late 1940s, when he wrote boiled down to: "obey the law, vote right, join some organizations and practice what conservation is profitable on your own land; the government will do the rest."(p.243-244)

Digitization

Currently the Digital Content Group of University of Wisconsin–Madison is conducting a large-scale digitization of Aldo Leopold's journals and records. They are expected to be made available online late 2009.[7]

See also

- Timeline of environmental events
- Land Ethic
- Sand County Foundation
- Yale School of Forestry & Environmental Studies
- Aldo Leopold Legacy Trail System

References

- Knight, Richard L. and Suzanne Riedel (ed). 2002. *Aldo Leopold and the Ecological Conscience*. Oxford University Press. ISBN 0195149440.
- Lorbiecki, Marybeth. 1996. *Aldo Leopold: A Fierce Green Fire*. Helena, Mont.: Falcon Press. ISBN 1560444789.
- McClintock, James I. 1994. *Nature's Kindred Spirits*. University of Wisconsin Press. ISBN 0299141748.
- Meine, Curt. 1988. *Aldo Leopold: His Life and Work*. Madison, Wis.: University of Wisconsin Press. ISBN 0299114902.
- Newton, Julianne Lutz. 2006. *Aldo Leopold's Odyssey*. Washington: Island Press/Shearwater Books. ISBN 9781597260459.

External links

- Aldo Leopold Foundation [8]
- Leopold Heritage Group [9]
- The Aldo Leopold Archives [10] Digitized archival materials held by the University of Wisconsin–Madison Archives.
- Leopold Conservation Award [11]
- Excerpts from the Works of Aldo Leopold [12]
- Works by or about Aldo Leopold [13] in libraries (WorldCat catalog)

References

[1] Errington, P.L. (1948) In Appreciation of Aldo Leopold. *The Journal of Wildlife Management*. 12(4) pp. 341-350

[2] Raitt, RJ (1984) In Memoriam: A. Starker Leopold. *Auk* 101: 868-871. PDF (http://elibrary.unm.edu/sora/Auk/v101n04/p0868-p0871. pdf)

[3] Mark Staves and Randy Wayne (2009) In Memoriam: A. Carl Leopold. *Lansing Star* Dec. 3, 2009. html obituary (http://www.lansingstar. com/content/view/5628/71/)

[4] Leopold, A. (1949) A Sand County Almanac (Ballantine Books ed., 1970)(p. 262)

[5] Leopold, Aldo Thinking like a Mountain (http://www.eco-action.org/dt/thinking.html)

[6] *Aldo Leopold: A Fierce Green Fire* By Marybeth Lorbiecki (Falcon Press, 1996), quote on back cover *Aldo Leopold: A Fierce Green Fire* (http://books.google.com/books?id=N8RZ7GqNSMUC&pg=PT1&lpg=PT1&dq=bruce+babbitt,++aldo+leopold&source=bl& ots=B6t4kzmivF&sig=do7fi4fS2UHHpsvSaz9fbThQ05s&hl=en&ei=vPP_Sr7-BoWKlAeC9MCPCw&sa=X&oi=book_result&ct=result& resnum=10&ved=0CCAQ6AEwCQ#v=onepage&q=bruce babbitt, aldo leopold&f=false)

[7] http://www.news.wisc.edu/15023

[8] http://www.aldoleopold.org/

[9] http://www.leopoldheritage.org/

[10] http://digital.library.wisc.edu/1711.dl/AldoLeopold

[11] http://www.leopoldconservationaward.org/

[12] http://gargravarr.cc.utexas.edu/chrisj/leopold-quotes.html

[13] http://worldcat.org/identities/lccn-n50-49888

A Sand County Almanac

Aldo Leopold	
Author	→ Aldo Leopold
Country	United States
Language	English
Subject(s)	→ Ecology, → Environmentalism
Publisher	Oxford University Press
Publication date	1949
Pages	240 pp
ISBN	0-19-500777-8
OCLC Number	3799061 [1]

A Sand County Almanac is a 1949 non-fiction book written by American ecologist and environmentalist → Aldo Leopold. Describing the land around Leopold's home in Sauk County, Wisconsin and his thoughts on developing a "land ethic," it was edited and published by his son, Luna, a year after Leopold's death from a heart attack. The collection of essays is considered to be a landmark book in the American → conservation movement.

The book has had over two million copies printed and has been translated into nine languages.[2] It has informed and changed the → environmental movement and stimulated a widespread interest in → ecology as a science. Google Scholar, for example, lists nearly 2700 citations of this book.

Overview

A Sand County Almanac is a combination of natural history, scene painting with words, and philosophy. It is perhaps best known for the following quote, which defines his land ethic: "A thing is right when it tends to preserve the integrity, stability, and beauty of the biotic community. It is wrong when it tends otherwise." The original publication format was issued by Oxford University Press in 1949. It incorporated a number of previously published essays that Leopold had been contributing to popular hunting and conservation magazines, along with a set of longer more philosophical essays. The final format was assembled by Luna Leopold shortly after his father's death, but based closely on notes that presumably reflected Aldo Leopold's intentions. Subsequent editions have changed both the format and the content of the essays included in the original.

In the original publishing, the book begins with a set of essays under the heading "Sand County Almanac," which is divided into twelve segments, one for each month. These essays mostly follow the changes in the ecology on Leopold's farm near Baraboo, Wisconsin. (There is, in fact, no "Sand County" in Wisconsin. The term "sand counties" refers to the a section of the state marked by sandy soils). There are anecdotes and observations about flora and fauna reactions to the seasons as well as mentions of conservation topics.

The second section of the book, "Sketches Here and There," shifts the rhetorical focus from time to place. The essays are thematically organized around farms and wildernesses in Canada, Mexico and the United States. Some of these essays are autobiographical. "Red Legs Kicking," for example, recounts Leopold's boyhood experience of hunting in Iowa. The seminal essay "Thinking Like a Mountain" recalls another hunting experience later in life that was formative for Leopold's later views. Here Leopold describes the death of she-wolf killed by his party during a time when conservationists were operating under the assumption that elimination of top predators would make game plentiful. The essay provides a non-technical characterization of the trophic cascade where the removal of single species carries serious implications for the rest of the ecosystem.[3]

The book ends with a section of philosophical essays grouped together under the heading "The Upshot". Here Leopold explores ironies of conservation: in order to promote wider appreciation of wild nature and engender necessary political support, one encourages recreational usage of wilderness that ultimately destroys it. Musings on "trophies" contrasts the way that some need a physical specimen to prove their conquest into the wilderness, though photographs may be less damaging that a trophy head to be mounted on the wall. He suggests that the best trophy is the experience of wilderness itself, along with its character building aspects. Leopold also rails against the way that policy makers need to find an economic motive for conservation. In the concluding essay, "A Land Ethic", Leopold delves into a more appropriate rationale for conservation. In "The Ecological Conscience" section. He wrote: "Conservation is a state of harmony between men and land." Leopold felt it was generally agreed that more conservation education was needed; however quantity and content were up for debate. He believed that land is not a commodity to be possessed; rather, humans must have mutual respect for Earth in order not to destroy it. He also puts forth the idea that humans will never be free if they have no wild spaces in which to roam.

Leopold's home, Aldo Leopold Shack and Farm, was listed on the U.S. National Register of Historic Places in 1978.

The book's importance and continuing influence

In a 1990 poll of the membership by the American Nature Study Society, *A Sand County Almanac* and *Silent Spring* stand alone as the two most venerated and impactful environmental books of the 20th century. [4] The book was little noticed when published but, during the environmental awakening of the 1970s, a paperback edition turned into a surprise bestseller. [5] It still sells about 40,000 copies a year.[6]

The book has had immense popular influence and has been described as: "one of the benchmark titles of the ecological movement"[7] , "a major influence on American attitudes toward our natural environment"[8] , "recognized as a classic piece of outdoor literature, rivalling Thoreau's Walden"[9] .

The book has also had great influence on environmental thinkers: "along with Walden and → Rachel Carson's Silent Spring, one of the main intellectual underpinnings of → environmentalism in America"[10] , "Leopold's essays set the tone for writing about conservation"[11] . Leopold, through his book, is cited as one of the founders of Deep Ecology.[12] The book has "attracted such overwhelming attention from environmental philosophers as a source of inspiration and ideas"[13] . Leopold himself has been described as: "a visionary who still influences American conservation policy"[14] .

See also

- Conservation in the United States

Further reading

- Knight, Richard L. and Suzanne Riedel. 2002. *Aldo Leopold and the Ecological Conscience*. Oxford University Press. ISBN 0195149440.

External links

- Excerpt Text on the "wolf killing" [15]

References

[1] http://worldcat.org/oclc/3799061

[2] The Aldo Leopold Foundation's site on the *Almanac* (http://www.aldoleopold.org/about/almanac.htm)

[3] Leopold, Aldo Thinking Like a Mountain (http://www.eco-action.org/dt/thinking.html)

[4] *"Silent Spring" and "A Sand County Almanac": The Two Most Significant Environmental Books of the 20th Century*, Nature Study, v44 n2-3 p6-8 Feb 1991

[5] Book review by Donella Meadows, director of the Sustainability Institute and an adjunct professor of environmental studies at Dartmouth College http://www.yesmagazine.org/article.asp?ID=956

[6] Book review http://web.mac.com/jaginsburg/germtales/Sand_County_Almanac.html

[7] Blog http://bfgb.wordpress.com/2008/09/23/a-sand-county-almanac-by-aldo-leopold/

[8] Nature Writing: The Tradition in English, Finch, Elder, p376

[9] Reflections on A Sand County Almanac by Don H. Meredith http://www.donmeredith.ca/outdoorsmen/SandCounty.html

[10] StoryLines midwest, David Long, http://74.125.93.132/search?q=cache:Vz__H-clAu4J:www.programminglibrarian.org/assets/files/dp/sand_county.pdf+%22a+sand+county+almanac%22+%22silent+spring%22&cd=6&hl=en&ct=clnk&gl=ca

[11] Blog http://bfgb.wordpress.com/2008/09/23/a-sand-county-almanac-by-aldo-leopold/

[12] http://en.wikipedia.org/wiki/Deep_ecology

[13] Rachel Carson: Legacy and Challenge, Callicott and Back

[14] Book review, http://forestry.about.com/cs/foresthistory1/gr/Aldo_asca.htm

[15] http://www.operationnoah.org/resources/inspirations/killing-wolf-famous-text-aldo-leopold-sand-county-almanac

Environmental Values

Environmental Values	
Abbreviated title(s)	EV
Discipline	Philosophy Economics
Language	English
Publication details	
Publisher	White Horse Press (United Kingdom)
Publication history	1992 - present
Frequency	4 issues per year
Indexing	
ISSN	0963-2719 [1]
Links	
• Journal homepage [2]	

Environmental Values (**EV**) is a peer-reviewed journal of → environmental ethics, → ecological economics and applied philosophy, published four times per year. Environmental Values brings together contributions from philosophy, economics, politics, sociology, geography, anthropology, ecology and other disciplines, which relate to the present and future environment of human beings and other species. In doing so it aims to clarify the relationship between practical policy issues and more fundamental underlying principles or assumptions.

Established in 1992 with editorial offices at the University of Lancaster, UK and edited by Alan Holland until 2007 when Clive L. Spash became editor-in-chief and editorial offices were moved to the University of Central Lancashire.

References

[1] http://www.worldcat.org/issn/0963-2719
[2] http://www.erica.demon.co.uk/EV.html

Anthropocentrism

Anthropocentrism (from Greek: ἄνθρωπος, *anthropos*, "human being"; and κέντρον, *kentron*, "center") is the belief that humans must be considered at the center of, and above any other aspect of, reality.[1] This concept is sometimes known as **humanocentrism** or **human supremacy**. It is especially strong in certain religious cultures, such as the common Protestant Christian translation of Genesis 1:26, which is taken to state that God gave man dominion over all other earthly creatures.[2] The current Latin Vulgate, the official Bible of the Catholic Christian church, as well as St. Jerome's original, lack this anthropocentric nature, instead saying that God holds man responsible for the care and fate of all earthly creatures.[3] [4]

Environmentalism

Anthropocentrism has been posited by some environmentalists, in such books as *Confessions of an Eco-Warrior* by Dave Foreman and *Green Rage* by Christopher Manes, as the underlying (if unstated) reason why humanity dominates and sees the need to "develop" most of the Earth. Anthropocentrism has been identified by these writers and others as a root cause of the → ecological crisis, human overpopulation, and the → extinctions of many non-human species.

Anthropocentrism, or human-centredness, is believed by some to be the central problematic concept in environmental philosophy, where it is used to draw attention to a systematic bias in traditional Western attitudes to the non-human world.[5] Val Plumwood has argued[6] [7] that anthropocentrism plays an analogous role in green theory to androcentrism in feminist theory and ethnocentrism in anti-racist theory. Plumwood calls human-centredness "anthrocentrism" to emphasise this parallel.

Defenders of anthropocentrist views point out that maintenance of a healthy, sustainable environment is necessary for human well-being as opposed for its own sake. The problem with a "shallow" viewpoint is not that it is human centered but that according to William Grey[8] "What's wrong with shallow views is not their concern about the well-being of humans, but that they do not really consider enough in what that well-being consists. According to this view, we need to develop an enriched, fortified anthropocentric notion of human interest to replace the dominant short-term, sectional and self-regarding conception."

One of the first extended philosophical essays addressing environmental ethics, John Passmore's *Man's Responsibility for Nature*[9] has been repeatedly criticised by defenders of → deep ecology because of its anthropocentrism, often claimed to be constitutive of traditional Western moral thought.[10]

Christianity

Some evangelical Christians have also been critical, viewing a human-centered worldview, rather than a Christ-centered or God-centered worldview, as a core societal problem. According to this viewpoint, humanity placing its own desires ahead of the teachings of The Bible leads to rampant selfishness and behavior viewed as sinful.

The use of the word "dominion" in *Genesis*, where it is written that God gives man dominion over all creatures, is controversial. Many Biblical scholars, especially Roman Catholic and other non-Protestant Christians, consider this to be a flawed translation of a word meaning "stewardship", which would indicate that mankind should take care of the earth and its various forms of life, but is not inherently better than any other form of life.[11]

In the 1985 CBC series "A Planet For the Taking", Dr. David Suzuki explored the Old Testament roots of anthropocentrism and how it shaped our view of non-human animals.

In his book *Pale Blue Dot*, author Dr. Carl Sagan also reflects on what he perceives to be the conceitedness and pettiness of anthropocentrism, specifically associating the doctrine with religious belief.[12]

Biocentrism

→ *Biocentrism* has been proposed as an antonym of anthropocentrism.[citation needed] It has also been proposed as a generalized form of anthropocentrism.[13]

In fiction

In science-fiction, *Humanocentrism* is the idea that humans, as both beings and a species, are the superior sentients. Essentially the equivalent of race supremacy on a galactic scale, it entails intolerant discrimination against sentient non-humans, much like race supremacists discriminate against those not of their race. This idea is countered by Anti-Humanism. At times, this ideal also includes fear of and superiority over robots, downplaying the ideas of cybernetic revolts and machine rule.

Humanocentrism is a central theme in the science-fiction comic book series *Nemesis the Warlock* in which humanity (here referred to as Terrans) have conquered much of the galaxy and seek to enslave all alien life. Humans are here depicted as antagonists, a somewhat (but not entirely) unusual plot device in science-fiction.

In the Star Wars universe, the Galactic Empire is shown to be humanocentric, ruthlessly subjugating alien worlds, enslaving many of them, and only employing humans in its military. Grand Admiral Thrawn is a notable exception to this rule, likely because of both his immense talent and his partially human bloodline.

In C. S. Lewis's *Prince Caspian*, the Telmarine invaders of Narnia attempt to wipe out the Talking Beasts, Dwarfs, and nature spirits. Some creatures go into hiding, and spirits of trees and rivers go into dormancy.

In J.R.R Tolkien's *The Lord of the Rings*, the deforestation around Saruman's tower and the attack of the ents as a result of it is a means of showing that men are not the centre of the universe and if they continue to treat nature as a secondary element of the world, nature will retaliate. [citation needed]

Further reading

- Bertalanffy, General System Theory (1993): 239-48
- → White, Lynn Townsend, Jr, "The Historical Roots of Our Ecologic Crisis [14]", *Science*, Vol 155 (Number 3767), March 10, 1967, pp 1203–1207
- Chew, Sing C. "Ecology in Command"
- Chew, Sing C. "Ecological Futures"

See also

- Anthropic principle
- Anthropocentric embodied energy analysis
- → Biocentrism
- Carbon chauvinism

- Chauvinism
- Deep Ecology
- → Ecocentrism
- Ecofeminism
- Existentialism

- Great ape personhood
- Gynocentrism
- Human exceptionalism
- Intrinsic value (animal ethics)

- Speciesism
- Technocentrism
- Theocentricism

References

[1] Anthropocentrism (http://www.merriam-webster.com/dictionary/anthropocentrism) - Merriam-Webster Dictionary.

[2] citation needed

[3] (http://www.sacred-texts.com/bib/vul/gen001.htm#026) - Genesis 1:26 (Original Latin Vulgate)

[4] (http://bibledbdata.org/onlinebibles/vulgate/01_001.htm) - Genesis 1:26 (Latin Vulgate as of 12/12/2009)

[5] Naess, A. 1973. 'The Shallow and the Deep, Long-Range Ecology Movement' *Inquiry* 16: 95-100

[6] Plumwood, V. 1993. *Feminism and the Mastery of Nature*. London: Routledge

[7] Plumwood, V. 1996. Androcentrism and Anthrocentrism: Parallels and Politics. *Ethics and the Environment* 1

[8] Grey, W. 1993. 'Anthropocentrism and Deep Ecology' *Australiasian Journal of Philosophy* 71: 463-475 (http://www.uq.edu.au/
~pdwgrey/pubs/anthropocentrism.html)

[9] Passmore, J. 1974. *Man's Responsibility for Nature* London: Duckworth

[10] Routley, R. and V. 1980. 'Human Chauvinism and Environmental Ethics' in *Environmental Philosophy* (eds) D.S. Mannison, M. McRobbie
and R. Routley. Canberra: ANU Research School of Social Sciences: 96-189

[11] (http://www.fourmilab.ch/etexts/www/Vulgate/Genesis.html)

[12] Carl Sagan - Pale Blue Dot (http://www.youtube.com/watch?v=shUTJAwWmsE)

[13] http://www.insurgentdesire.org.uk/biocentrism.htm

[14] http://www.asa3.org/ASA/PSCF/1969/JASA6-69White.html

Biocentric individualism

'Biocentric Individualism' is a system of → environmental ethics proposed by noted environmental ethicist Dr. Gary Varner. It is, in part, a revision of the mental state theory of individual welfare, asserting that there is a hierarchy of things of moral importance:

- **Ground project**: Things that answer the question "Why is life worth living?" and consume a significant portion of an individual's life.
- **Non-biological interests**: Interests that aren't as important as ground projects, but more important than mere biological needs.
- **Biological needs**: The lowest classification of needs that are still worthy of moral consideration.

See also

- Maslow's hierarchy of needs

Biocentrism

For the politico-ecological concept, see Biocentrism (ethics).

Biocentrism and **biocentric** (from Greek: βίος, *bios*, "life"; and κέντρον, *kentron*, "center") are terms implying a centrality of life, nature, or biology. According to biocentrism, life creates the universe rather than the other way around. In this view, current theories of the physical world do not work, and can never be made to work, until they fully account for life and consciousness.

A biocentric theory proposed by American scientist Robert Lanza builds on quantum physics.[1] [2] His theory places biology before the other sciences in an attempt to solve one of nature's biggest puzzles: the *theory of everything* that other disciplines have been pursuing for the last century.[3] [4] [5] Lanza argues that biocentrism is falsifiable, and that future experiments, such as scaled-up quantum superposition, will either support or contradict the theory.[6]

Theory

The central claim of biocentrism is that what we call space and time are forms of animal sense perception, rather than external physical objects.[7]

Lanza argues that biocentrism offers insight into several major puzzles of science, including Heisenberg's uncertainty principle and the double-slit experiment, and into the forces, constants, and laws that shape the universe as we perceive it.[2]

According to a *Discover* magazine article adapted from Lanza's book, "biocentrism offers a more promising way to bring together all of physics, as scientists have been trying to do since Einstein's unsuccessful unified field theories of eight decades ago."[8]

Lanza's theory of biocentrism has seven principles:[9]

1. What we perceive as reality is a process that involves our consciousness. An "external" reality, if it existed, would by definition have to exist in space. But this is meaningless, because space and time are not absolute realities but rather tools of the human and animal mind.
2. Our external and internal perceptions are inextricably intertwined. They are different sides of the same coin and cannot be divorced from one another.
3. The behavior of subatomic particles, indeed all particles and objects, is inextricably linked to the presence of an observer. Without the presence of a conscious observer, they at best exist in an undetermined state of probability waves.
4. Without consciousness, "matter" dwells in an undetermined state of probability. Any universe that could have preceded consciousness only existed in a probability state.
5. The structure of the universe is explainable only through biocentrism. The universe is fine-tuned for life, which makes perfect sense as life creates the universe, not the other way around. The "universe" is simply the complete spatio-temporal logic of the self.
6. Time does not have a real existence outside of animal-sense perception. It is the process by which we perceive changes in the universe.
7. Space, like time, is not an object or a thing. Space is another form of our animal understanding and does not have an independent reality. We carry space and time around with us like turtles with shells. Thus, there is no absolute self-existing matrix in which physical events occur independent of life.

Reception

Lanza's article and book on "biocentrism" have received a mixed reception.

David Thompson, an astrophysicist at NASA's Goddard Space Flight Center, said that Lanza's "work is a wake-up call."[10] Nobel laureate (in Physiology or Medicine) E. Donnall Thomas said, "Any short statement does not do justice to such a scholarly work. The work is a scholarly consideration of science and philosophy that brings biology into the central role in unifying the whole."[11] Arizona State University physicist Lawrence Krauss stated, "It may represent interesting philosophy, but it doesn't look, at first glance, as if it will change anything about science."[12] Wake Forest University scientist Anthony Atala stated, "This new theory is certain to revolutionize our concepts of the laws of nature for centuries to come."[13] In *USA Today Online*, astrophysicist and science writer David Lindley asserted that Lanza's concept was a "vague, inarticulate metaphor" and stated that "I certainly don't see how thinking his way would lead you into any new sort of scientific or philosophical insight. That's all very nice, I would say to Lanza, but now what? I [also] take issue with his views about physics."[14] Daniel Dennett, a Tufts University philosopher, said he did not think the concept meets the standard of a philosophical theory. "It looks like an opposite of a theory, because he doesn't explain how it [consciousness] happens at all. He's stopping where the fun begins."[15] Richard Conn Henry, Professor of Physics and Astronomy at Johns Hopkins University, pointed out that Lanza's theory is consistent with quantum mechanics: "What Lanza says in this book is not new. Then why does Robert have to say it at all? It is because we, the physicists, do NOT say it—or if we do say it, we only whisper it, and in private—furiously blushing as we mouth the words. True, yes; politically correct, hell no!"[16] Indian physician and writer[17] [18] Deepak Chopra stated that "Lanza's insights into the nature of consciousness [are] original and exciting" and that "his theory of biocentrism is consistent with the most ancient wisdom traditions of the world which says that consciousness conceives, governs, and becomes a physical world. It is the ground of our Being in which both subjective and objective reality come into existence."[19]

See also

- Solipsism
- Subjective universe

References

[1] Alan Boyle (2007-03-08). " Theory of Every-Living-Thing (http://cosmiclog.msnbc.msn.com/archive/2007/03/08/85328.aspx)". *Cosmic Log*. msnbc.com. . Retrieved 2009-03-15.

[2] Lanza, Robert and Berman, Bob (2009). *Biocentrism: How Life and Consciousness are the Keys to Understanding the True Nature of the Universe*. BenBella. ISBN 978-1933771694.

[3] Aaron Rowe (2007-03-08). " Will Biology Solve the Universe? (http://www.wired.com/medtech/genetics/news/2007/03/72910)". *Wired*. . Retrieved 2009-03-15.

[4] " Biocentrism (http://www.the-scientist.com/news/display/55621/)". *The Scientist*. . Retrieved 2009-04-17.

[5] "A New Theory of the Universe" (http://www.theamericanscholar.org/a-new-theory-of-the-universe/), Spring 2007 *The American Scholar*

[6] Eric Berger (2009-08-23). " Book Spotlight: Biocentrism (http://blogs.chron.com/sciguy/archives/2009/08/book_spotlight_biocentrism. html)". *Houston Chronicle Blogs*. . Retrieved 2009-12-10.

[7] "Biocentrism: How life creates the universe" (http://www.msnbc.msn.com/id/31393080/), *MSNBC.com*

[8] "The Biocentric Universe Theory" (http://discovermagazine.com/2009/may/01-the-biocentric-universe-life-creates-time-space-cosmos/ article_view?b_start:int=1&-C=/), May 2009 "Discover magazine"

[9] Robert Lanza, MD with Bob Berman. " Biocentrism: How Life and Consciousness Are the Keys to Understanding the True Nature of the Universe (http://www.robertlanza.com/ biocentrism-how-life-and-consciousness-are-the-keys-to-understanding-the-true-nature-of-the-universe/)". .

[10] " Spring2009_Working01.qxd (http://www.moxiestudio.com/designsamples/BenBellaBooks_Spring2009_Catalog_Final.pdf)" (PDF). . Retrieved 2009-08-17.

[11] " A Biotech Provocateur Takes On Physics - Forbes.com (http://www.forbes.com/2007/03/09/ lanza-theories-physics-biotech-oped-cx_mh_0309lanza.html?partner=yahootix)". Forbes.com<!. . Retrieved 2009-08-09.

[12] "A Biotech Provocateur Takes On Physics" (http://www.forbes.com/2007/03/09/ lanza-theories-physics-biotech-oped-cx_mh_0309lanza.html?partner=yahootix), *Forbes.com*, 9 March 2007

[13] " Biocentrism: How Life and Consciousness Are the Keys to Understanding the True Nature of the Universe (9781933771694): Robert
      Lanza, Bob Berman: Books (http://www.amazon.com/gp/product/product-description/1933771690/ref=dp_proddesc_0?ie=UTF8&
      n=283155&s=books)". Amazon.com. . Retrieved 2009-08-17.

[14] " Exclusive: Response to Robert Lanza's essay (http://www.usatoday.com/tech/science/2007-03-09-lanza-response_N.htm)".
      Usatoday.Com. 2007-03-09. . Retrieved 2009-08-17.

[15] " A Biotech Provocateur Takes On Physics (http://www.forbes.com/2007/03/09/
      lanza-theories-physics-biotech-oped-cx_mh_0309lanza.html?partner=yahootix)". Forbes.com. . Retrieved 2009-08-17.

[16] " lanza (http://henry.pha.jhu.edu/biocentrism.pdf)" (PDF). . Retrieved 2009-08-17.

[17] " We have the power to change the world for the better (http://www.un.org/special-rep/ohrlls/ohrlls/Statements/hr message 01 Dec
      04_world-peace_puerto rico-details.pdf?eventID=421&action=eventDetails)". United Nations. 2004-12-01. . Retrieved 2009-08-22. ""...
      New Humanity Forum ... founding members include ...; Indian physician and writer Deepak Chopra;""

[18] " CNN LIVE EVENT/SPECIAL: CNN International Simulcast: Terrorist Attacks in India (Rush transcript) (http://transcripts.cnn.com/
      TRANSCRIPTS/0811/27/se.04.html)". cnn.com. 2008-11-27. . Retrieved 2009-08-22. ""Our Jonathan Mann posed that question to the
      Deepak Chopra, the Indian physician and philosopher... DEEPAK CHOPRA, CHOPRA CENTER FOR WELLBEING""

[19] " Dr. Robert Lanza is Featured Guest on Deepak Chopra's SIRIUS XM Stars Radio Show (http://ca.news.finance.yahoo.com/s/
      13082009/31/link-f-prnewswire-dr-robert-lanza-featured-guest-deepak-chopra-s.html/)". Yahoo Finance. . Retrieved 2009-12-15.

Conservation (ethic)

For the laws of conservation in the physical sciences, see conservation law.

Conservation is an ethic of resource use, allocation, and protection. Its primary focus is upon maintaining the health of the natural world: its forests, fisheries, habitats, and biological diversity. Secondary focus is on materials conservation and energy conservation, which are seen as important to protect the natural world. Those who follow the conservation ethic and, especially, those who advocate or work toward conservation goals are termed conservationists.

Satellite photograph of deforestation in progress in the Tierras Bajas
project in eastern Bolivia. Photograph courtesy NASA.

Introduction

To conserve habitat in terrestrial ecoregions and stop deforestation is a goal widely shared by many groups with a wide variety of motivations. *These issues and groups are covered in their own articles.*

To protect sea life from → extinction due to overfishing is another commonly stated goal of conservation — ensuring that "some will be available for our children" to continue a way of life.

The consumer conservation ethic is sometimes expressed by the *four R's*: " Rethink, Reduce, Reuse, Recycle," This social ethic primarily relates to local purchasing, moral purchasing, the sustained, and efficient use of renewable resources, the moderation of destructive use of finite resources, and the prevention of harm to common resources such as air and water quality, the natural functions of a living earth, and cultural values in a built environment.

Much attention has been given to preserving the natural characteristics of Hopetoun Falls, Australia, while allowing access for visitors.

The principal value underlying most expressions of the conservation ethic is that the natural world has intrinsic and intangible worth along with utilitarian value — a view carried forward by the scientific → conservation movement and some of the older Romantic schools of ecology movement.

More Utilitarian schools of conservation seek a proper valuation of local and global impacts of human activity upon nature in their effect upon human well being, now and to our posterity. How such values are assessed and exchanged among people determines the social, political, and personal restraints and imperatives by which conservation is practiced. This is a view common in the modern → environmental movement.

These movements have diverged but they have deep and common roots in the → conservation movement.

In the United States of America, the year 1864 saw the publication of two books which laid the foundation for Romantic and Utilitarian conservation traditions in America. The posthumous publication of Henry David Thoreau's *Walden* established the grandeur of unspoiled nature as a citadel to nourish the spirit of man. From George Perkins Marsh a very different book, *Man and Nature*, later subtitled "The Earth as Modified by Human Action", catalogued his observations of man exhausting and altering the land from which his sustenance derives.

Terminology

> The conservation of natural resources is the fundamental problem. Unless we solve that problem, it will avail us little to solve all others.
>
> —Theodore Roosevelt[1]

In common usage, the term refers to the activity of systematically protecting natural resources such as forests, including biological diversity. Carl F. Jordan defines the term as:[2]

> biological **conservation** as being a philosophy of managing the environment in a manner that does not despoil, exhaust or extinguish.

While this usage is not new, the idea of biological conservation has been applied to the principles of → ecology, biogeography, anthropology, economy and sociology to maintain biodiversity.

The term "conservation" itself may cover the concepts such as cultural diversity, genetic diversity and the concept of movements environmental conservation, seedbank (preservation of seeds). These are often summarized as the priority to respect diversity, especially by Greens.

Much recent movement in conservation can be considered a resistance to commercialism and globalization. Slow food is a consequence of rejecting these as moral priorities, and embracing a slower and more locally-focused lifestyle.

Practice

Distinct trends exist regarding conservation development. While many countries' efforts to preserve species and their habitats have been government-led, those in the North Western Europe tended to arise out of the middle-class and aristocratic interest in natural history, expressed at the level of the individual and the national, regional or local learned society. Thus countries like Britain, the Netherlands, Germany, etc. had what we would today term NGOs — in the shape of the RSPB, National Trust and County Naturalists' Trusts (dating back to 1889, 1895 and 1912 respectively) Natuurmonumenten, Provincial conservation Trusts for each Dutch province, Vogelbescherming, etc. — a long time

Bachalpsee in the Swiss Alps; generally mountainous areas are less affected by human activity.

before there were National Parks and National Nature Reserves. This in part reflects the absence of wilderness areas in heavily cultivated Europe, as well as a longstanding interest in laissez-faire government in some countries, like the UK, leaving it as no coincidence that John Muir, the British-born founder of the National Park movement (and hence of government-sponsored conservation) did his sterling work in the USA, where he was the motor force behind the establishment of such NPs as Yosemite and Yellowstone. Nowadays, officially more than 10 percent of the world is legally protected in some way or the other, and in practice private fundraising is insufficient to pay for the effective management of so much land with protective status.

The Daintree Rainforest in Queensland, Australia.

Protected areas in developing countries, where probably as many as 70-80 percent of the species of the world live, still enjoy very little effective management and protection. Although some countries such as Mexico have non-profit civil organizations and land owners dedicated to protect vast private property, such is the case of Hacienda Chichen's Maya Jungle Reserve and Bird Refuge [3] in Chichen Itza, Yucatán. The Adopt A Ranger Foundation has calculated that worldwide about 140,000 rangers are needed for the protected areas in developing and transition countries. There are no data on how many rangers are employed at the moment, but probably less than half the protected areas in developing and transition countries have any rangers at all and those that have them are at least 50% short This means that there would be a worldwide ranger deficit of 105,000 rangers in the developing and transition countries.

One of the world's foremost conservationists, Dr. Kenton Miller, stated about the importance of rangers: "The future of our ecosystem services and our heritage depends upon park rangers. With the rapidity at which the challenges to protected areas are both changing and increasing, there has never been more of a need for well prepared human capacity to manage. Park rangers are the backbone of park management. They are on the ground. They work on the front line with scientists, visitors, and members of local communities."

Adopt A Ranger,[4] fears that the ranger deficit is the greatest single limiting factor in effectively conserving nature in 75% of the world. Currently, no conservation organization or western country or international organization addresses this problem. Adopt A Ranger has been incorporated to draw worldwide public attention to the most urgent problem that conservation is facing in developing and transition countries: protected areas without field staff. Very specifically, it will contribute to solving the problem by fund raising to finance rangers in the field. It will also help governments in developing and transition countries to assess realistic staffing needs and staffing strategies

See also

- Conservation biology
- → Conservation movement
- → Ecology
- → Environmentalism
- Environmental protection
- Habitat conservation
- List of conservation issues
- List of environmental issues
- List of environmental organizations
- List of environmental topics
- Natural environment
- Natural capital
- Natural resource
- Renewable resource
- Sustainability
- Sustainable agriculture
- Water conservation

References

[1] Theodore Roosevelt, Address to the Deep Waterway Convention Memphis, TN, October 4, 1907

[2] Jordan, Carl (1995). *Replacing Quantity With Quality As a Goal for Global Management.* Wiley. ISBN 0471595152.

[3] Hacienda Chichen The Importance of Eco-Design (http://www.haciendachichen.com/eco-design.htm)

[4] http://www.adopt-a-ranger.org

- Conservation and evolution (Frankel et Soulé, 1981)
- Glacken, C.J. (1967) Traces on the Rhodian Shore. University of California Press. Berkeley
- Grove, R.H. (1992) 'Origins of Western Environmentalism', Scientific American 267(1): 22-27.
- Grove, R.H. (1997) *Ecology, Climate and Empire: Colonialism and Global Environmental History 1400-1940* Cambridge: Whitehorse Press
- Grove, R.H. (1995) *Green Imperialism: Colonial Expansion, Tropical Island Edens, and the Origins of Environmentalism, 1600-1860* New York: Cambridge University Press
- Leopold, A. (1966) *A Sand County Almanac* New York: Oxford University Press
- Pinchot, G. (1901) *The Fight for Conservation* New York: Harcourt Brace.
- *"Why Care for Earth's Environment?"* (in the series *"The Bible's Viewpoint"*) is a two-page article in the December 2007 issue of the magazine Awake!.

External links

- *Dictionary of the History of ideas*: (http://etext.lib.virginia.edu/cgi-local/DHI/dhi.cgi?id=dv1-59) Conservation of Natural Resources
- *For Future Generations*, a Canadian documentary on how the conservation ethic influenced national parks (http://beta.nfb.ca/film/For_Future_Generations/)

Conservation movement

The **conservation of forests,** also known as **nature conservation,** is a political and social movement that seeks to protect natural resources including plant and animal species as well as their habitat for the future.

The early conservation movement included fisheries and wildlife management, water, soil conservation and sustainable forestry. The contemporary conservation movement has broadened from the early movement's emphasis on use of sustainable yield of natural resources and preservation of wilderness areas to include preservation of biodiversity. Some say the conservation movement is part of the broader and more far-reaching → environmental movement, while others argue that they differ both in ideology and practice. Chiefly in the United States, conservation is seen as differing from →

The High Peaks Wilderness Area in the 6000000-acre (24000 km²) Adirondack Park is a publicly-protected area located in northeast New York.

environmentalism in that it aims to preserve natural resources expressly for their continued sustainable use by humans.[1] In other parts of the world conservation is used more broadly to include the setting aside of natural areas and the active protection of wildlife for their inherent value, as much as for any value they may have for humans.

Much attention has been given to preserving the natural characteristics of Hopetoun Falls, Australia, while allowing ample access for visitors.

History

The nascent conservation movement slowly developed in the 19th century, starting first in the scientific forestry methods pioneered by the Germans and the French in the 17th and 18th centuries. While continental Europe created the scientific methods later used in conservationist efforts, British India and the United States are credited with starting the conservation movement.

Foresters in India, often German, managed forests using early climate change theories (in America, see also, George Perkins Marsh) that Alexander von Humboldt developed in the mid 19th century, applied fire protection, and tried to keep the "house-hold" of nature. This was an early ecological idea, in order to preserve the growth of delicate teak trees. The same German foresters who headed the Forest Service of India, such as Dietrich Brandis and Berthold Ribbentrop, traveled back to Europe and taught at

F. V. Hayden's map of Yellowstone National Park, 1871.

forestry schools in England (Cooper's Hill, later moved to Oxford). These men brought with them the legislative and scientific knowledge of conservationism in British India back to Europe, where they distributed it to men such as Gifford Pinchot, which in turn helped bring European and British Indian methods to the United States.

Areas of concern

Deforestation and overpopulation are issues affecting all regions of the world. The consequent destruction of wildlife habitat has prompted the creation of conservation groups in other countries, some founded by local hunters who have witnessed declining wildlife populations first hand. Also, it was highly important for the conservation movement to solve problems of living conditions in the cities and the overpopulation of such places.

Boreal forest and the Arctic

The idea of incentive conservation is a modern one but its practice has clearly defended some of the sub Arctic wildernesses and the wildlife in those regions for thousands of years, especially by indigenous peoples such as the Evenk, Yakut, Sami, Inuit and Cree. The fur trade and hunting by these peoples have preserved these regions for thousands of years. Ironically, the pressure now upon them comes from non-renewable resources such as oil, sometimes to make synthetic clothing which is advocated as a humane substitute for fur. (See Raccoon Dog for case study of the conservation of an animal through fur trade.) Similarly, in the case of the beaver, hunting and fur trade were thought to bring about the animal's demise, when in fact they were an integral part of its conservation. For many years children's books stated and still do, that the decline in the beaver population was due to the fur trade. In reality however, the decline in beaver numbers was because of habitat destruction and deforestation, as well as its continued persecution as a pest (it causes flooding). In Cree lands however, where the population valued the animal for meat and fur, it continued to thrive. The Inuit defend their relationship with the seal in response to outside critics.[2]

Latin America (Bolivia)

The Izoceño-Guaraní of Santa Cruz, Bolivia is a tribe of hunters who were influential in establishing the Capitania del Alto y Bajo Isoso (CABI). CABI promotes economic growth and survival of the Izoceno people while discouraging the rapid destruction of habitat within Bolivia's Gran Chaco. They are responsible for the creation of the 34,000 square kilometre Kaa-Iya del Gran Chaco National Park and Integrated Management Area (KINP). The KINP protects the most biodiverse portion of the Gran Chaco, an ecoregion shared with Argentina, Paraguay and Brazil. In 1996, the Wildlife Conservation Society joined forces with CABI to institute wildlife and hunting monitoring programs in 23 Izoceño communities. The partnership combines traditional beliefs and local knowledge with the political and administrative tools needed to effectively manage habitats. The programs rely solely on voluntary participation by local hunters who perform self-monitoring techniques and keep records of their hunts. The information obtained by the hunters participating in the program has provided CABI with important data required to make educated decisions about the use of the land. Hunters have been willing participants in this program because of pride in their traditional activities, encouragement by their communities and expectations of benefits to the area.

Africa (Botswana)

In order to discourage illegal South African hunting parties and ensure future local use and sustainability, indigenous hunters in Botswana began lobbying for and implementing conservation practices in the 1960s. The Fauna Preservation Society of Ngamiland (FPS) was formed in 1962 by the husband and wife team: Robert Kay and June Kay, environmentalists working in conjunction with the Batawana tribes to preserve wildlife habitat.

The FPS promotes habitat conservation and provides local education for preservation of wildlife. Conservation initiatives were met with strong opposition from the Botswana government because of the monies tied to big-game hunting. In 1963, BaTawanga Chiefs and tribal hunter/adventurers in conjunction with the FPS founded Moremi National Park and Wildlife Refuge, the first area to be set aside by tribal people rather than governmental forces. Moremi National Park is home to a variety of wildlife, including lions, giraffes, elephants, buffalo, zebra, cheetahs and antelope, and covers an area of 3,000 square kilometers. Most of the groups involved with establishing this protected land were involved with hunting and were motivated by their personal observations of declining wildlife and habitat.

See also

- Conservation biology
- Conservation ethic
- → Ecology
- Ecology movement
- → Environmental movement
- Environmental protection
- → Environmentalism
- Forest protection
- Habitat conservation
- List of environmental organizations
- List of environment topics
- Natural environment
- The Evolution of the Conservation Movement, 1850–1920
- Water conservation
- Wildlife conservation
- Wildlife management

Further reading

- Bates, J. Leonard. "Fulfilling American Democracy: The Conservation Movement, 1907 to 1921", *The Mississippi Valley Historical Review,* Vol. 44, No. 1. (Jun., 1957), pp. 29–57. in JSTOR [3]
- Barton, Gregory A. *Empire Forestry and the Origins of Environmentalism,* Cambridge University Press, 2001
- Bolaane, Maitseo. "Chiefs, Hunters & Adventurers: The Foundation of the Okavango/Moremi National Park, Botswana". Journal of Historical Geography. 31.2 (Apr. 2005): 241-259.
- Clover, Charles. 2004. *The End of the Line: How overfishing is changing the world and what we eat.* Ebury Press, London. ISBN 0-09-189780-7
- Hays, Samuel P. "Conservation and the Gospel of Efficiency" Harvard University Press, 1959.
- Herring, Hall and Thomas McIntyre. "Hunting's New Ambassadors (Sporting Conservation Council) ". Field and Stream. 111.2 (June 2006): p. 18.
- Judd, Richard W. "Common Lands and Common People, The Origins of Conservation in Northern New England" Harvard University Press, 1997
- Nash, Roderick. "Wilderness and the American Mind" Yale University Press, 1967
- Noss, Andrew and Imke Oetting. "Hunter Self-Monitoring by the Izoceño -Guarani in the Bolivian Chaco". Biodiversity & Conservation. 14.11 (2005): 2679-2693.
- Pope, Carl. "A Sporting Chance – Sportsmen and Sportswomen are some of the biggest supporters for the preservation of wildlife". Sierra. 81.3 (May/June 1996): 14.
- Reiger, George. "Common Ground: Battles Over Hunting Only Draw Attention Away From the Real Threat to Wildlife". Field and Stream. 100.2 (June 1985): p. 12.
- Reiger, George. "Sportsmen Get No Respect (Media Ignores Role of Sportsmen in Conservation) ". Field and Stream. 101.10 (Feb 1997): p. 18.

External links

- A Lot To Say [4] – Spreading Environmentally Positive Messages Through Organic Clothing
- The Everglades in the Time of Marjorie Stoneman Douglas [5] Photo exhibit created by the State Archives of Florida
- British Virgin Islands Conservation Coral Reef Disaster Documentary [6] This documentary exposes the challenges facing the Islanders who are battling the government to curtail this development disaster.
- A history of conservation in New Zealand [7]
- *For Future Generations*, a Canadian documentary on conservation and national parks [8]
- Amazon Rainforest Fund [9]

References

[1] Gifford, John C. (1945). *Living by the Land*. Coral Gables, Florida: Glade House. pp. 8. ASIN B0006EUXGQ (http://www.amazon.com/dp/B0006EUXGQ).

[2] " Inuit Ask Europeans to Support Its Seal Hunt and Way of Life (http://www.icc.gl/UserFiles/File/sealskin/2006-03-07_icc_saelskind_pressemeddelse_eng.pdf)". 6 March 2006. . Retrieved 12 July 2007.

[3] http://links.jstor.org/sici?sici=0161-391X%28195706%2944%3A1%3C29%3AFADTCM%3E2.0.CO%3B2-O

[4] http://shop.alottosay.com

[5] http://www.floridamemory.com/PhotographicCollection/photo_exhibits/everglades.cfm

[6] http://www.documentary-film.net/search/video-listings.php?e=81

[7] http://www.teara.govt.nz/TheBush/Conservation/ConservationAHistory/en

[8] http://beta.nfb.ca/film/For_Future_Generations/

[9] http://amazonrainforestfund.org

Deep ecology

Part of the Politics series on

Green politics

🌀 **Environment Portal**
Politics portal

Deep ecology is a somewhat recent branch of ecological philosophy (ecosophy) that considers humankind as an integral part of its environment. The philosophy emphasizes the equal value of human and non-human life as well as the importance of the ecosystem and natural processes. It provides a foundation for the → environmental and green movements and has led to a new system of → environmental ethics.

The retreat of Aletsch Glacier in the Swiss Alps (situation in 1979, 1991 and 2002).

Deep ecology's core principle is the claim that, like humanity, the living environment as a whole has the same right to live and flourish. Deep ecology describes itself as "deep" because it persists in asking deeper questions concerning "why" and "how" and thus is concerned with the fundamental philosophical questions about the impacts of human life as one part of the ecosphere, rather than with a narrow view of → ecology as a branch of biological science, and aims to avoid merely anthropocentric environmentalism, which is concerned with conservation of the environment only for exploitation by and for humans purposes, which excludes the fundamental philosophy of deep ecology. Deep ecology seeks a more holistic view of the world we live in and seeks to apply to life the understanding that separate parts of the ecosystem (including humans) function as a whole.

Development

The phrase "deep ecology" was coined by the Norwegian philosopher Arne Næss in 1973,[1] and he helped give it a theoretical foundation. "For Arne Næss, ecological science, concerned with facts and logic alone, cannot answer ethical questions about how we should live. For this we need ecological wisdom. Deep ecology seeks to develop this by focusing on deep experience, deep questioning and deep commitment. These constitute an interconnected system. Each gives rise to and supports the other, whilst the entire system is, what Næss would call, an ecosophy: an evolving but consistent philosophy of being, thinking and acting in the world, that embodies ecological wisdom and harmony."[2] Næss rejected the idea that beings can be ranked according to their relative value. For example, judgments on whether an animal has an eternal soul, whether it uses reason or whether it has consciousness (or indeed higher consciousness) have all been used to justify the ranking of the human animal as superior to other animals. Næss states that from an ecological point of view "the right of all forms [of life] to live is a universal right which cannot be quantified. No single species of living being has more of this particular right to live and unfold than any other species." This metaphysical idea is elucidated in Warwick Fox's claim that we and all other beings are "aspects of a single unfolding reality".[3] . As such Deep Ecology would support the view of → Aldo Leopold in his book, → *A Sand County Almanac* that humans are "plain members of the biotic community". They also would support Leopold's "Land Ethic": "a thing is right when it tends to preserve the integrity, stability and beauty of the biotic community. It is wrong when it tends otherwise." Daniel Quinn in *Ishmael*, showed that an anthropocentric myth underlies our current view of the world, and a jellyfish would have an equivalent jellyfish centric view[4] .

Deep ecology offers a philosophical basis for environmental advocacy which may, in turn, guide human activity against perceived self-destruction. Deep ecology and → environmentalism hold that the science of ecology shows that ecosystems can absorb only limited change by humans or other dissonant influences. Further, both hold that the actions of modern civilization threaten global ecological well-being. Ecologists have described change and stability in ecological systems in various ways, including homeostasis, dynamic equilibrium, and "flux of nature".[5] Regardless of which model is most accurate, environmentalists contend that massive human economic activity has pushed the biosphere far from its "natural" state through reduction of biodiversity, climate change, and other influences. As a consequence, civilization is causing mass extinction. Deep ecologists hope to influence social and political change through their philosophy.

Scientific

Næss and Fox do not claim to use logic or induction to derive the philosophy directly from scientific ecology [6] but rather hold that scientific ecology directly implies the metaphysics of deep ecology, including its ideas about the self and further, that deep ecology finds scientific underpinnings in the fields of → ecology and system dynamics.

In their 1985 book *Deep Ecology*,[7] Bill Devall and George Sessions describe a series of sources of deep ecology. They include the science of ecology itself, and cite its major contribution as the rediscovery in a modern context that "everything is connected to everything else". They point out that some ecologists and natural historians, in addition to their scientific viewpoint, have developed a deep ecological consciousness—for some a political consciousness and at times a spiritual consciousness. This is a perspective beyond the strictly human viewpoint, beyond → anthropocentrism. Among the scientists they mention particularly are → Rachel Carson, → Aldo Leopold, John Livingston, Paul R. Ehrlich and Barry Commoner, together with Frank Fraser Darling, Charles Sutherland Elton, Eugene Odum and Paul Sears.

A further scientific source for deep ecology adduced by Devall and Sessions is the "new physics." which they describe as shattering Descartes's and Newton's vision of the universe as a machine explainable in terms of simple linear cause and effect, and instead providing a view of Nature in constant flux and the idea that observers are separate an illusion. They refer to Fritjof Capra's *The Tao of Physics* and *The Turning Point* for their characterisation of how the new physics leads to metaphysical and ecological views of interrelatedness, which, according to Capra, should make deep ecology a framework for future human societies. Devall and Sessions also credit the American poet and social critic Gary Snyder—with his devotion to Buddhism, Native American studies, the outdoors, and alternative social movements—as a major voice of wisdom in the evolution of their ideas.

The scientific version of the Gaia hypothesis was also an influence on the development of deep ecology.

Spiritual

The central spiritual tenet of deep ecology is that the human species is a part of the Earth and not separate from it. A process of self-realisation or "re-earthing" is used for an individual to intuitively gain an ecocentric perspective. The notion is based on the idea that the more we *expand the self* to identify with "others" (people, animals, ecosystems), the more we realize ourselves. Transpersonal psychology has been used by Warwick Fox to support this idea.

In relation to the Judeo-Christian tradition, Næss offers the following criticism: "The arrogance of stewardship [as found in the Bible] consists in the idea of superiority which underlies the thought that we exist to watch over nature like a highly respected middleman between the Creator and Creation."[8] This theme had been expounded in → Lynn Townsend White, Jr.'s 1967 article "The Historical Roots of Our Ecological Crisis",[9] in which however he also offered as an alternative Christian view of man's relation to nature that of Saint Francis of Assisi, who he says spoke for the equality of all creatures, in place of the idea of man's domination over creation.

Experiential

Drawing upon the Buddhist tradition is the work of Joanna Macy. Macy, working as an anti-nuclear activist in the USA, found that one of the major impediments confronting the activists' cause was the presence of unresolved emotions of despair, grief, sorrow, anger and rage. The denial of these emotions led to apathy and disempowerment.

We may have intellectual understanding of our interconnectedness, but our culture, experiential deep ecologists like John Seed argue, robs us of emotional and visceral experience of that interconnectedness which we had as small children, but which has been socialised out of us by a highly anthropocentric alienating culture.

Through "Despair and Empowerment Work" and more recently "The Work that Reconnects", Macy and others have been taking Experiential Deep Ecology into many countries including especially the USA, Europe (particularly Britain and Germany), Russia and Australia.

Philosophical – Spinoza and deep ecology

Arne Næss, who first wrote about the idea of deep ecology, from the early days of developing this outlook conceived Spinoza as a philosophical source.[10]

Others have followed Naess' inquiry, including Eccy de Jonge, in *Spinoza and Deep Ecology: Challenging Traditional Approaches to Environmentalism* [11], and Brenden MacDonald, in *Spinoza, Deep Ecology, and Human Diversity—Realization of Eco-Literacies* [12]

One of the topical centres of inquiry connecting Spinoza to Deep Ecology is "self-realization." See Arne Naess in *The Shallow and the Deep, Long-Range Ecology movement* [13] and *Spinoza and the Deep Ecology Movement* [14] for discussion on the role of Spinoza's conception of self-realization and its link to deep ecology.

Principles

Proponents of deep ecology believe that the world does not exist as a resource to be freely exploited by humans. The ethics of deep ecology hold that a whole system is superior to any of its parts. They offer an eight-tier platform to elucidate their claims:[15]

1. The well-being and flourishing of human and nonhuman life on Earth have value in themselves (synonyms: intrinsic value, inherent value). These values are independent of the usefulness of the nonhuman world for human purposes.
2. Richness and diversity of life forms contribute to the realization of these values and are also values in themselves.
3. Humans have no right to reduce this richness and diversity except to satisfy vital human needs.
4. The flourishing of human life and cultures is compatible with a substantial decrease of the human population. The flourishing of nonhuman life requires such a decrease.
5. Present human interference with the nonhuman world is excessive, and the situation is rapidly worsening.
6. Policies must therefore be changed. These policies affect basic economic, technological, and ideological structures. The resulting state of affairs will be deeply different from the present.
7. The ideological change is mainly that of appreciating life quality (dwelling in situations of inherent value) rather than adhering to an increasingly higher standard of living. There will be a profound awareness of the difference between big and great.
8. Those who subscribe to the foregoing points have an obligation directly or indirectly to try to implement the necessary changes.

Movement

In practice, deep ecologists support decentralization, the creation of ecoregions, the breakdown of industrialism in its current form, and an end to authoritarianism.

Deep ecology is not normally considered a distinct movement, but as part of the green movement. The deep ecological movement could be defined as those within the green movement who hold deep ecological views. Deep ecologists welcome the labels "Gaian" and "Green" (including the broader political implications of this term, e.g. commitment to peace). Deep ecology has had a broad general influence on the green movement by providing an independent ethical platform for Green parties, political ecologists and environmentalists.

The philosophy of deep ecology helped differentiate the modern ecology movement by pointing out the anthropocentric bias of the term "→ environment", and rejecting the idea of humans as authoritarian guardians of the environment.

Criticism

Interests in nature

Animal rights activists state that for something to require rights and protection intrinsically, it must have interests.[16] Deep ecology is criticised for assuming that plants, for example, have their own interests as they are manifested by the plant's behavior - self-preservation being considered an expression of a will to live, for instance. Deep ecologists claim to *identify* with non-human nature, and in doing so, criticise those who claim they have no understanding of what non-human nature's desires and interests are. The criticism is that the interests that a deep ecologist purports to give to non-human organisms such as survival, reproduction, growth and prosperity are really human interests. "The earth is endowed with 'wisdom', wilderness equates with 'freedom', and life forms are said to emit 'moral' qualities."[17] It has also been argued that species and ecosystems themselves have rights.[18] However, the overarching criticism assumes that humans, in governing their own affairs, are somehow immune from this same assumption; i.e. how can governing humans truly presume to understand the interests of the rest of humanity. While the deep ecologist critic would answer that the logical application of language and social mores would provide this justification, i.e. voting patterns etc, the deep ecologist would note that these "interests" are ultimately observable solely from the logical application of the behavior of the life form, which is the same standard used by deep ecologists to perceive the standard of interests for the natural world.

Deepness

Deep ecology is criticised for its claim to be *deeper* than alternative theories, which by implication are *shallow*. However despite repeated complaints about use of the term it still enjoys wide currency; *deep* evidently has an attractive resonance for many who seek to establish a new ethical framework for guiding human action with respect to the natural world. It may be presumptuous to assert that one's thinking is deeper than others'. When Arne Næss coined the term *deep ecology* he compared it favourably with *shallow environmentalism* which he criticized for its utilitarian and anthropocentric attitude to nature and for its materialist and consumer-oriented outlook.[19] [20] Against this is Arne Næss's own view that the "depth" of deep ecology resides in the persistence of its interrogative questioning, particularly in asking "Why?" when faced with initial answers.

Ecofeminist response

Both ecofeminism and deep ecology put forward a new conceptualization of the self. Some ecofeminists, such as Marti Kheel,[21] argue that self-realization and identification with all nature places too much emphasis on the whole, at the expense of the independent being. Ecofeminists contend that their concept of the self (as a dynamic process consisting of relations) is superior. Ecofeminists would also place more emphasis on the problem of androcentrism rather than → anthropocentrism.

Misunderstanding scientific information

Daniel Botkin[22] has likened deep ecology to its antithesis, the wise use movement, when he says that they both "misunderstand scientific information and then arrive at conclusions based on their misunderstanding, which are in turn used as justification for their ideologies. Both begin with an ideology and are political and social in focus." Elsewhere though, he asserts that deep ecology must be taken seriously in the debate about the relationship between humans and nature because it challenges the fundamental assumptions of western philosophy. Botkin has also criticized Næss's restatement and reliance upon the balance of nature idea and the perceived contradiction between his argument that all species are morally equal and his disparaging description of pioneering species.

"Shallow" View superior

Writer William Grey believes that developing a non-anthropocentric set of values is "a hopeless quest" He seeks an improved "shallow" view, writing, "What's wrong with shallow views is not their concern about the well-being of humans, but that they do not really consider enough in what that well-being consists. We need to develop an enriched, fortified anthropocentric notion of human interest to replace the dominant short-term, sectional and self-regarding conception."[23]

Deep ecology as not "deep" enough

Social ecologists such as Murray Bookchin[24] claim that deep ecology fails to link environmental crises with authoritarianism and hierarchy. Social ecologists believe that environmental problems are firmly rooted in the manner of human social interaction, and protest that an ecologically sustainable society could still be socially exploitative. Deep ecologists reject the argument that ecological behavior is rooted in the social paradigm (according to their view, that is an anthropocentric fallacy), and they maintain that the converse of the social ecologists' objection is also true in that it is equally possible for a socially egalitarian society to continue to exploit the Earth.

Links with other movements

Parallels have been drawn between deep ecology and other movements, in particular the animal rights movement and Earth First!.

Peter Singer's 1975 book *Animal Liberation* critiqued anthropocentrism and put the case for animals to be given moral consideration. This can be seen as a part of a process of expanding the prevailing system of ethics to wider groupings. However, Singer has disagreed with deep ecology's belief in the intrinsic value of nature separate from questions of suffering, taking a more utilitarian stance. The feminist and civil rights movements also brought about expansion of the ethical system for their particular domains. Likewise deep ecology brought the whole of nature under moral consideration.[25] The links with animal rights are perhaps the strongest, as "proponents of such ideas argue that 'All life has intrinsic value'".[26]

Many in the radical environmental direct-action movement Earth First! claim to follow deep ecology, as indicated by one of their slogans *No compromise in defence of mother earth*. In particular, David Foreman, the co-founder of the movement, has also been a strong advocate for deep ecology, and engaged in a public debate with Murray Bookchin on the subject.[27] [28] Judi Bari was another prominent Earth Firster who espoused deep ecology. Many Earth First! actions have a distinct deep ecological theme; often these actions will be to save an area of old growth forest, the habitat of a snail or an owl, even individual trees. It should however be noted that, especially in the United Kingdom, there are also strong anti-capitalist and anarchist currents in the movement, and actions are often symbolic or have other political aims. At one point Arne Næss also engaged in environmental direct action, though not under the Earth First! banner, when he tied himself to a Norwegian fjord in a successful protest against the building of a dam.[29]

Robert Greenway and Theodore Roszak have employed the Deep Ecology (DE) platform as a means to argue for Ecopsychology. Although Ecopsychology is a highly differentiated umbrella that encompasses many practices and perspectives, its ethos is generally consistent with DE. As this now almost forty-year old "field" expands and continues to be reinterpreted by a variety of practitioners, social and natural scientists, and humanists, "ecopsychology" may change to include these novel perspectives.

Early Influences

- Mary Hunter Austin | Ralph Waldo Emerson | → Aldo Leopold
- John Muir | Henry David Thoreau

Notable advocates of deep ecology

- David Abram
- Judi Bari
- Thomas Berry
- Wendell Berry
- Leonardo Boff
- Fritjof Capra
- Bill McKibben
- Michael Dowd
- David Foreman

- Vivienne Elanta
- Warwick Fox
- Edward Goldsmith
- Felix Guattari
- Martin Heidegger (controversial: see Development above)
- Derrick Jensen
- Dolores LaChapelle
- Pentti Linkola (controversial)
- John Livingston
- Paul Hawken
- Joanna Macy

- Jerry Mander
- Freya Mathews
- Terence McKenna
- W.S. Merwin
- Arne Næss
- David Orton
- Daniel Quinn
- Theodore Roszak
- Savitri Devi (controversial)
- John Seed
- Paul Shepard
- Gary Snyder
- Richard Sylvan
- Douglas Tompkins

- Oberon Zell-Ravenheart
- John Zerzan

See also

- Anarcho-primitivism
- Coupled human-environment system
- Earth liberation
- EcoCommunalism
- Ecopsychology
- Environmental psychology
- EcoTheology
- Gaia hypothesis
- Growth Fetish
- Human ecology
- Neotribalism
- Negative Population Growth | Population Connection
- Pathetic fallacy
- Permaculture
- Systems theory | The Great Story
- Sustainable development
- The Revenge of Gaia
- Voluntary Human Extinction Movement

Notes

[1] Næss, Arne (1973) 'The Shallow and the Deep, Long-Range Ecology Movement.' Inquiry 16: 95-100

[2] Harding, Stephan (2002), "What is Deep Ecology"

[3] Fox, Warwick, (1990) *Towards a Transpersonal Ecology* (Shambhala Books)

[4] Quinn, Daniel (1995), "Ishmael: An Adventure of the Mind and Spirit" (Bantam)

[5] Botkin, Daniel B. (1990). *Discordant Harmonies: A New Ecology for the Twenty-First Century*. Oxford Univ. Press, NY, NY. ISBN 0-19-507469-6.

[6] : The Shallow and the Deep, Long Range Ecology movements A summary by Arne Naess (http://www.alamut.com/subj/ideologies/ pessimism/Naess_deepEcology.html)

[7] Devall, Bill; Sessions, George (1985). *Deep Ecology*. Gibbs M. Smith. ISBN 0-87905-247-3. pp. 85-88

[8] Næss, Arne. (1989). *Ecology, Community and Lifestyle: Outline of an Ecosophy*. p. 187. ISBN 0-521-34873-0

[9] White, Jr, Lynn Townsend (March 1967). "The Historical Roots of Our Ecological Crisis". *Science* 155 (3767): 1203–1207. doi: 10.1126/science.155.3767.1203 (http://dx.doi.org/10.1126/science.155.3767.1203). PMID 17847526 (http://www.ncbi.nlm.nih.gov/ pubmed/17847526). (HTML copy (http://www.zbi.ee/~kalevi/lwhite.htm), PDF copy (http://web.lemoyne.edu/~glennon/ LynnWhitearticle.pdf)).

[10] Spinoza and Deep Ecology (http://www.springerlink.com/content/e8213222t8hk5736/)

[11] http://ndpr.nd.edu/review.cfm?id=2601

[12] http://www.newciv.org/mem/prof-newslog.php?did=373&vid=373&xmode=show_article&artid=000373-000019&amode=standard& aoffset=0&time=1246755640

[13] http://books.google.ca/books?id=HTBMPKH9_2UC&source=gbs_navlinks_s

[14] http://www.springerlink.com/content/x36131180168g245/

[15] Devall and Sessions, *op. cit.*, p. 70.

[16] Feinberg, Joel. " The Rights of Animals and Future Generations (http://www.animal-rights-library.com/texts-m/feinberg01.htm)". . Retrieved 2006-04-25.

[17] Joff (2000). " The Possibility of an Anti-Humanist Anarchism (http://library.nothingness.org/articles/anar/en/display/310)". . Retrieved 2006-04-25.

[18] Pister, E. Phil (1995). " The Rights of Species and Ecosystems (http://www.nativefish.org/articles/Fish_Rights.php)". *Fisheries* 20 (4). . Retrieved 2006-04-25.

[19] Great River Earth Institute. " Deep Ecology: Environmentalism as if all beings mattered (http://www.greatriv.org/de.htm)". . Retrieved 2006-04-25.

[20] Panaman, Ben. " Animal Ethics Encyclopedia: Deep Ecology (http://www.animalethics.org.uk/aec-d-entries.html#Deep Ecology)". . Retrieved 2006-04-25.

[21] Kheel, Marti. (1990): *Ecofeminism and Deep Ecology; reflections on identity and difference* from: Diamond, Irene. Orenstein. Gloria (editors), *Reweaving the World; The emergence of ecofeminism*. Sierra Club Books. San Francisco. pp 128-137. ISBN 0-87156-623-0

[22] Botkin, Daniel B. (2000). *No Man's Garden: Thoreau and a New Vision for Civilization and Nature*. Shearwater Books. pp. 42, 39. ISBN 1-55963-465-0.

[23] Anthropocentrism and Deep Ecology by William Grey (http://www.uq.edu.au/~pdwgrey/pubs/anthropocentrism.html)

[24] Bookchin, Murray (1987). " Social Ecology versus Deep Ecology: A Challenge for the Ecology Movement (http://dwardmac.pitzer.edu/ Anarchist_Archives/bookchin/socecovdeepeco.html)". *Green Perspectives/Anarchy Archives*. .

[25] Alan AtKisson. " Introduction To Deep Ecology, an interview with Michael E. Zimmerman (http://www.context.org/ICLIB/IC22/ Zimmrman.htm)". *In Context* (22). . Retrieved 2006-05-04.

[26] Wall, Derek (1994). *Green History*. Routledge. ISBN 0-415-07925-X.

[27] David Levine, ed (1991). *Defending the Earth: a dialogue between Murray Bookchin and Dave Foreman*.

[28] Bookchin, Murray; Graham Purchace, Brian Morris, Rodney Aitchtey, Robert Hart, Chris Wilbert (1993). *Deep Ecology and Anarchism*. Freedom Press. ISBN 0-900384-67-0.

[29] J. Seed, J. Macy, P. Flemming, A. Næss, *Thinking like a mountain: towards a council of all beings*, Heretic Books (1988), ISBN 0-946097-26-7, ISBN 0-86571-133-X.

Caution by the link to "The Shallow and the Deep" - there are several faults in the quote of the original article. (Added words, wrong commas which can by misleading)

Bibliography

- Bender, F. L. 2003. *The Culture of Extinction: Toward a Philosophy of Deep Ecology* Amherst, New York: Humanity Books.
- Devall, W. and G. Sessions. 1985. *Deep Ecology: Living As if Nature Mattered* Salt Lake City: Gibbs M. Smith, Inc.
- Drengson, Alan. 1995. *The Deep Ecology Movement*
- Katz, E., A. Light, et al. 2000. *Beneath the Surface: Critical Essays in the Philosophy of Deep Ecology* Cambridge, Mass.: MIT Press.
- LaChapelle, D. 1992. *Sacred Land, Sacred Sex: Rapture of the Deep* Durango: Kivakí Press.
- Næss, A. 1989. *Ecology, Community and Lifestyle: Outline of an Ecosophy* Translated by D. Rothenberg. Cambridge: Cambridge University Press.
- Nelson, C. 2006. *Ecofeminism vs. Deep Ecology*, Dialogue, San Antonio, TX: Saint Mary's University Dept. of Philosophy.
- Passmore, J. 1974. *Man's Responsibility for Nature* London: Duckworth.
- Sessions, G. (ed) 1995. *Deep Ecology for the Twenty-first Century* Boston: Shambhala.
- Taylor, B. and M. Zimmerman. 2005. *Deep Ecology" in B. Taylor, ed., Encyclopedia of Religion and Nature, v 1, pp. 456-60, London: Continuum International. Also online at* (http://www.religionandnature.com/ern/sample. htm)

Further reading

- David Abram, *The Spell of the Sensuous: Perception and Language in a More-than-Human World.* (1996) Pantheon Books.
- Conesa-Sevilla, J. (2006). The Intrinsic Value of the Whole: Cognitive and utilitarian evaluative processes as they pertain to ecocentric, deep ecological, and ecopsychological "valuing." *The Trumpeter*, 22, 2, 26-42.
- Jozef Keulartz, *Struggle for nature : a critique of radical ecology*, London [etc.] : Routledge, 1998
- Michael Tobias ed, *Deep Ecology*, Avant Books (1984, 1988) ISBN 0-932238-13-0.
- Carolyn Merchant, *The Death of Nature*, HarperOne (1990) ISBN 0062505955, 978-0062505958.
- Harold Glasser (ed), *The Selected Works of Arne Næss*, Volumes 1-10. Springer, (2005), ISBN 1-4020-3727-9. (review (http://home.ca.inter.net/~greenweb/Naess_Appreciation.html))
- Jack Turner, *The Abstract Wild*, Tucson, Univ of Arizona Press (1996)
- de Steiguer, J.E. 2006. *The Origins of Modern Environmental Thought.* The University of Arizona Press. 246 pp.

Educational Programs

- Naropa University Master of Arts Transpersonal Psychology, Ecopsychology Concentration (http://www. naropa.edu/academics/graduate/psychology/tcp/ecoc/)

External links

- Northwest Earth Institute (http://www.nwei.org) Discussion course on Deep Ecology
- Philosophy, Cosmology and Conciousness, California Institude of Integral Studies. (http://www.ciis.edu/pcc/)
- Downloadable interview with Dr. Alan Drengson about Deep Ecology and Arne Næss. June 6, 2008. (http:// besustainable.com/greenmajority/2008/06/06/tgm-88/)
- Nature Worship in Hinduism (http://www.hinduwisdom.info/Nature_Worship.htm)
- Church of Deep Ecology (http://www.churchofdeepecology.org/)
- Deep Ecology Movement (http://www.deepecology.org/movement.htm), Alan Drengson, Foundation for Deep Ecology.

- Environmental Ethics Journal (http://www.cep.unt.edu/enethics.html)
- Foundation for Deep Ecology (http://www.deepecology.org/)
- Green Parties World Wide (http://www.greens.org/)
- The Great Story (http://www.thegreatstory.org/) - a leading Deep Ecology/Deep Time educational website
- Gaia Foundation (http://gaia.iinet.net.au): an Australian organisation based upon the principles of Deep Ecology. See especially its links page.
- The Green Web (http://home.ca.inter.net/~greenweb/index.htm) a left biocentric environmental research group, with a number of writings on deep ecology
- The Trumpeter (http://trumpeter.athabascau.ca/), Canadian journal of ecosophy, quite a number of articles from Næss among others
- Welcome to All Beings (http://www.joannamacy.net): Joanna Macy on the work of Experiential Deep Ecology
- Social Ecology vs Deep Ecology (http://dwardmac.pitzer.edu/ANARCHIST_ARCHIVES/bookchin/socecovdeepeco.html) - A Challenge for the Ecology Movement by Murray Bookchin
- Deep Ecology in the Song of Songs (http://www.song-of-songs.net)

Ecocentrism

Ecocentrism (from Greek: οἶκος, oikos, "house"; and κέντρον, kentron, "center". Pronounced ekō'sen‚trizəm) is a term used in ecological political philosophy to denote a nature-centred, as opposed to human-centred, system of values. The justification for ecocentrism usually consists in an ontological belief and subsequent ethical claim. The ontological belief denies that there are any existential divisions between human and non-human nature sufficient to claim that humans are either (a) the sole bearers of intrinsic value or (b) possess greater intrinsic value than non-human nature. Thus the subsequent ethical claim is for an equality of intrinsic value across human and non-human nature, or 'biospherical egalitarianism'.[1] According to Rowe:[2]

> The ecocentric argument is grounded in the belief that, compared to the undoubted importance of the human part, the whole ecosphere is even more significant and consequential: more inclusive, more complex, more integrated, more creative, more beautiful, more mysterious, and older than time. The "environment" that → anthropocentrism misperceives as materials designed to be used exclusively by humans, to serve the needs of humanity, is in the profoundest sense humanity's source and support: its ingenious, inventive life-giving matrix. Ecocentrism goes beyond → biocentrism with its fixation on organisms, for in the ecocentric view people are inseparable from the inorganic/organic nature that encapsulates them. They are particles and waves, body and spirit, in the context of Earth's ambient energy.[3]

and:

> To switch Western culture from its present track to a saving ecopolitical route means finding a new and compelling belief-system to redirect our way-of-living. It must be a vital outgrowth from our science-based culture. It seems to me that the only promising universal belief-system is ecocentrism, defined as a value-shift from Homo sapiens to planet earth. A scientific rationale backs the value-shift. All organisms are evolved from Earth, sustained by Earth. Thus Earth, not organism, is the metaphor for Life. Earth not humanity is the Life-center, the creativity-center. Earth is the whole of which we are subservient parts. Such a fundamental philosophy gives ecological awareness and sensitivity an enfolding, material focus.

> Ecocentrism is not an argument that all organisms have equivalent value. It is not an anti-human argument nor a put-down of those seeking social justice. It does not deny that myriad important homocentric problems exist. But it stands aside from these smaller, short-term issues in order to consider Ecological Reality. Reflecting on the ecological status of all organisms, it comprehends the

Ecosphere as a Being that transcends in importance any one single species, even the self-named sapient one.[4]

Origin of term

The ecocentric ethic was conceived by → Aldo Leopold[5] and recognizes that all species, including humans, are the product of a long evolutionary process and are inter-related in their life processes.[6] The writings of → Aldo Leopold and his idea of the land ethic and good environmental management are a key element to this philosophy. Ecocentrism focuses on the biotic community as a whole and strives to maintain ecosystem composition and ecological processes.[7] The term also finds expression in the first principle of the → deep ecology movement, as formulated by Arne Naess and George Sessions in 1984[8] which points out that → anthropocentrism, which considers humans as the center of the universe and the pinnacle of all creation, is a difficult opponent for ecocentrism.[9]

Background

Environmental thought and the various branches of the environmental movement are often classified into two intellectual camps: those that are considered anthropocentric, or "human-centred," in orientation and those considered biocentric, or "life-centred." This division has been described in other terminology as "shallow" ecology versus "deep" ecology and as "technocentrism" versus "ecocentrism". Ecocentrism can be seen as one stream of thought within → environmentalism, the political and ethical movement that seeks to protect and improve the quality of the natural environment through changes to environmentally harmful human activities by adopting environmentally benign forms of political, economic, and social organization and through a reassessment of humanity's relationship with nature. In various ways, environmentalism claims that non-human organisms and the natural environment as a whole deserve consideration when appraising the morality of political, economic, and social policies.[10]

Relationship to other similar philosophies

Anthropocentrism

Ecocentrism is taken by its proponents to constitute a radical challenge to long-standing and deeply rooted anthropocentric attitudes in Western culture, science, and politics. Anthropocentrism is alleged to leave the case for the protection of non-human nature subject to the demands of human utility, and thus never more than contingent on the demands of human welfare. An ecocentric ethic, by contrast, is believed to be necessary in order to develop a non-contingent basis for protecting the natural world. Critics of ecocentrism have argued that it opens the doors to an anti-humanist morality that risks sacrificing human well-being for the sake of an ill-defined 'greater good'.[11] → Deep ecologist Arne Naess has identified anthropocentrism as a root cause of the → ecological crisis, human overpopulation, and the → extinctions of many non-human species.[12] Others point to the gradual historical realization that humans are not the centre of all things, that "A few hundred years ago, with some reluctance, Western people admitted that the planets, Sun and stars did not circle around their abode. In short, our thoughts and concepts though irreducibly anthropomorphic need not be anthropocentric."[13]

Technocentrism

Ecocentrism is also contrasted with technocentrism (meaning values centred on technology) as two opposing perspectives on attitudes towards human technology and its ability to affect, control and even protect the environment. Ecocentrics, including "deep green" ecologists, see themselves as being subject to nature, rather than in control of it. They lack faith in modern technology and the bureaucracy attached to it. Ecocentrics will argue that the natural world should be respected for its processes and products, and that low impact technology and self-reliance is more desirable than technological control of nature.[14] Technocentrics, including imperialists, have absolute faith in technology and industry and firmly believe that humans have control over nature. Although technocentrics may accept that environmental problems do exist, they do not see them as problems to be solved by a reduction in industry. Rather, environmental problems are seen as problems to be solved using science. Indeed, technocentrics see that the way forward for developed and developing countries and the solutions to our environmental problems today lie in scientific and technological advancement.[14]

Biocentrism

The distinction between biocentrism and ecocentrism is ill-defined. However, biocentrism is discussed in detail by Lanza and Berman.[15] Ecocentrism recognizes Earths interactive living and non-living systems rather than just the Earth's organisms (biocentrism) as central in importance.[16]

A Manifesto for Earth

In 2004 Ted Mosquin[17] and Stan Rowe published *A Manifesto for Earth* in the journal Biodiversity. This was a synthesis of their deliberations on ecocentrism and contained a Statement of Conviction followed by a set of Core Principles with their associated Action Principles.[18] These are reproduced here:

Statement of conviction

Everyone searches for meaning in life, for supportive convictions that take various forms. Many look to faiths that ignore or discount the importance of this world, not realizing in any profound sense that we are born from Earth and sustained by it throughout our lives. In today's dominating industrial culture, Earth-as-home is not a self-evident percept. Few pause daily to consider with a sense of wonder the enveloping matrix from which we came and to which, at the end, we all return. Because we are issue of the Earth, the harmonies of its lands, seas, skies and its countless beautiful organisms carry rich meanings barely understood. We are convinced that until the Ecosphere is recognized as the indispensable common ground of all human activities, people will continue to set their immediate interests first. Without an ecocentric perspective that anchors values and purposes in a greater reality than our own species, the resolution of political, economic, and religious conflicts will be impossible. Until the narrow focus on human communities is broadened to include Earth's ecosystems — the local and regional places wherein we dwell — programs for healthy sustainable ways of living will fail. A trusting attachment to the Ecosphere, an aesthetic empathy with surrounding Nature, a feeling of awe for the miracle of the Living Earth and its mysterious harmonies, is humanity's largely unrecognized heritage. Affectionately realized again, our connections with the natural world will begin to fill the gap in lives lived in the industrialized world. Important ecological purposes that civilization and urbanization have obscured will re-emerge. The goal is restoration of Earth's diversity and beauty, with our prodigal species once again a cooperative, responsible, ethical member.

Core principles

Principle 1 The Ecosphere is the Center of Value for Humanity

Principle 2 The Creativity and Productivity of Earth's Ecosystems Depend on their Integrity

Principle 3 The Earth-centered Worldview is supported by Natural History

Principle 4 Ecocentric Ethics are Grounded in Awareness of our Place in Nature

Principle 5 An Ecocentric Worldview Values Diversity of Ecosystems and Cultures

Principle 6 Ecocentric Ethics Support Social Justice

Action principles

Principle 7 Defend and Preserve Earth's Creative Potential

Principle 8 Reduce Human Population Size

Principle 9 Reduce Human Consumption of Earth Parts

Principle 10 Promote Ecocentric Governance

Principle 11 Spread the Message

Ecocentrism and the law

An ecocentric ethic can assist in forming a basis for traditional legal methods that are deficient in dealing with many environmental problems. For example, national laws cannot resolve international problems, and can place undue reliance on economic measures as a surrogate for ecological values.[19]

See also

- → Biocentrism
- → Deep ecology
- Earth liberation
- Ecocentric embodied energy analysis
- → Environmentalism

- Ecological humanities
- Ecofeminism
- Gaia hypothesis
- Holocentric
- Technocentrism

Further reading

- Bosselmann, K. 1999. *When Two Worlds Collide: Society and Ecology*. ISBN 0-9597948-3-2
- Eckersley, R. 1992. *Environmentalism and Political Theory: Toward an Ecocentric Approach*. State University of New York Press.
- Hettinger, N. and Throop, B. 1999. Refocusing Ecocentrism. *Environmental Ethics* 21: 3-21.

External links

- Ecospheric Ethics [20]
- Marginalization of Ecocentrism [21]

References

[1] Political Dictionary. (http://www.answers.com/topic/ecocentrismAnswers.com) Accessed 13 June 2009.

[2] Stan Rowe, biographic profile. (http://www.ecospherics.net/pages/aboutauthors.html#rowe)

[3] Rowe, Stan J. (1994). "Ecocentrism: the Chord that Harmonizes Humans and Earth." (http://www.ecospherics.net/pages/RoweEcocentrism.html) *The Trumpeter* **11**(2): 106-107.

[4] Rowe Stan J. "Ecocentrism and Traditional Ecological Knowledge." (http://www.ecospherics.net/pages/Ro993tek_1.html)

[5] Leopold, A. 1949. *A sand county almanac.* New York: Oxford University Press.

[6] Lindenmeyer, D. & Burgman, M. 2005. *Practical conservation biology.* CSIRO Publishing, Collingwood, Australia. ISBN 0-643-09089-4

[7] Booth, D.E. 1992. The economics and ethics of old growth forests. *Environmental Ethics* 14: 43-62.

[8] Arne Naess|Naess, Arne & Sessions, George 1984. "A Deep Ecology Eight Point Platform" cited in *Deep Ecology for the 21st Century*, Readings on the Philosophy and Practice of the New Environmentalism, ed. George Sessions, Shambhala, Boston and London, 1995.

[9] "Ecocentrism and the Deep Ecology Platform". (http://www.academon.com/lib/paper/102996.html)

[10] " environmentalism (http://www.britannica.com/EBchecked/topic/765710/environmentalism)". *Encyclopædia Britannica*. 2009. . Retrieved 13 June 2009.

[11] Ecocentrism at answers.com (http://www.answers.com/topic/ecocentrism)

[12] Naess, Arne 1973. "The Shallow and the Deep, Long-Range Ecology Movement". *Inquiry* **16**: 95-100

[13] see Rowe (http://www.ecospherics.net/pages/RoweEcocentrism.html)

[14] "Earth, ecocentrism and Technocentrism". (http://www.sustainable-environment.org.uk/Earth/Ecocentrism_and_Technocentrism.php)

[15] Lanza, Robert and Berman, Bob (2009). *Biocentrism: How Life and Consciousness are the Keys to Understanding the True Nature of the Universe.* BenBella. ISBN 978-1933771694.

[16] "Ecocentrism". The Oxford Pocket Dictionary of Current English. 2009. Encyclopedia.com. (http://www.encyclopedia.com) Accessed 13 June 2009.

[17] (http://www.ecospherics.net/pages/aboutauthors.html#rowe) Ted Mosquin, biographic profile.

[18] Mosquin, T. & Rowe, S. (2004). "Manifesto for Earth." (http://www.ecospherics.net/pages/EarthManifesto.pdf) *Biodiversity* **5**(1): 3-9.

[19] Tolan, P. 2009. Ecocentric Perspectives on Global Warming: Toward an Earth Jurisprudence. (http://gsj.cgpublisher.com/about.html) *Global Studies Journal* **1**(4): 39-50.

[20] http://www.ecospherics.net/

[21] http://techlobyte.tripod.com/warez/paper4222006.html

Environmental movement

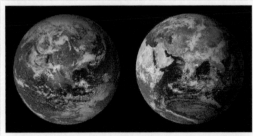

Blue Marble composite images generated by NASA in 2001 (left) and 2002 (right).

> The conservation of natural resources is the fundamental problem. Unless we solve that problem, it will avail us little to solve all others.
>
> —Theodore Roosevelt[1]

The **environmental movement**, a term that includes the → conservation and green movements, is a diverse scientific, social, and political movement for addressing environmental issues.

→ Environmentalists advocate the sustainable management of resources and stewardship of the environment through changes in public policy and individual behavior. In its recognition of humanity as a participant in (not enemy of) ecosystems, the movement is centered on → ecology, health, and human rights.

The environmental movement is represented by a range of organizations, from the large to grassroots. Due to its large membership, varying and strong beliefs, and occasionally speculative nature, the environmental movement is not always united in its goals. At its broadest, the movement includes private citizens, professionals, religious devotees, politicians, and extremists.

Introduction

The roots of the modern environmental movement can be traced to attempts in nineteenth-century Europe and North America to expose the costs of environmental negligence, notably disease, as well as widespread air and water pollution, but only after the Second World War did a wider awareness begin to emerge.

The US environmental movement emerged in the late nineteenth and early twentieth century, with two key strands: protectionists such as John Muir wanted land and nature set aside for its own sake, while conservationists such as Gifford Pinchot wanted to manage natural resources for exploitation. Among the early protectionists that stood out as leaders in the movement were Henry David Thoreau, John Muir and George Perkins Marsh. Thoreau was concerned about the wildlife in Massachusetts; he wrote *Walden; or, Life in the Woods* as he studied the wildlife from a cabin. John Muir founded the Sierra Club, one of the largest conservation organizations in the United States. Marsh was influential with regards to the need for resource conservation. Muir was instrumental in the creation of the world's first national park at Yellowstone in 1872.

During the 1950s, 1960s, and 1970s, several events illustrated the magnitude of environmental damage caused by humans. In 1954, the 23 man crew of the Japanese fishing vessel *Lucky Dragon 5* was exposed to radioactive fallout from a hydrogen bomb test at Bikini Atoll. The publication of the book *Silent Spring* (1962) by → Rachel Carson drew attention to the impact of chemicals on the natural environment. In 1967, the oil tanker *Torrey Canyon* went aground off the southwest coast of England, and in 1969 oil spilled from an offshore well in California's Santa Barbara Channel. In 1971, the conclusion of a law suit in Japan drew international attention to the effects of decades of mercury poisoning on the people of Minamata.[2]

At the same time, emerging scientific research drew new attention to existing and hypothetical threats to the environment and humanity. Among them were Paul R. Ehrlich, whose book *The Population Bomb* (1968) revived concerns about the impact of exponential population growth. Biologist Barry Commoner generated a debate about growth, affluence and "flawed technology." Additionally, an association of scientists and political leaders known as the Club of Rome published their report *The Limits to Growth* in 1972, and drew attention to the growing pressure on natural resources from human activities.

Meanwhile, technological accomplishments such as nuclear proliferation and photos of the Earth from outer space provided both new insights and new reasons for concern over Earth's seemingly small and unique place in the universe.

In 1972, the United Nations Conference on the Human Environment was held in Stockholm, and for the first time united the representatives of multiple governments in discussion relating to the state of the global environment. This conference led directly to the creation of government environmental agencies and the UN Environment Program. The United States also passed new legislation such as the Clean Water Act, the Clean Air Act, the Endangered Species Act, and the National Environmental Policy Act- the foundations for current environmental standards.

Since the 1970s, public awareness, environmental sciences, → ecology, and technology have advanced to include modern focus points like ozone depletion, global climate change, acid rain, and the potentially harmful genetically modified organisms (GMOs).

Scope of the movement

Biological studies

- Environmental science is the study of the interactions among the physical, chemical and biological components of the environment.
- → Ecology, or ecological science, is the scientific study of the distribution and abundance of living organisms and how these properties are affected by interactions between the organisms and their environment.

Primary focus points

Before flue gas desulfurization was installed, the air-polluting emissions from this power plant in New Mexico contained excessive amounts of sulfur dioxide.

- The environmental movement is broad in scope and can include any topic related to the environment, conservation, and biology, as well as preservation of landscapes, flora, and fauna for a variety of purposes and uses. See List of environmental issues. When an act of violence is committed against someone or some institution in the name of environmental defense it is referred to as eco terrorism
- The → Conservation movement seeks to protect natural areas for sustainable consumption, as well as traditional (hunting, fishing, trapping) and spiritual use.

Other focus points

- Environmental Conservation is the process in which one is involved in conserving the natural aspects of the environment. Whether through reforestation, recycling, or pollution control, environmental conservation sustains the natural quality of life.
- Environmental health movement dates at least to Progressive Era, and focuses on urban standards like clean water, efficient sewage handling, and stable population growth. Environmental health could also deal with nutrition, preventive medicine, aging, and other concerns specific to human well-being. Environmental health is also seen as an indicator for the state of the environment, or an early warning system for what may happen to humans.
- Environmental Justice is a movement that began in the U.S. in the 1980s and seeks an end to environmental racism and prevent low-income and minority communities from an unbalanced exposure to highways, garbage dumps, and factories. The Environmental Justice movement seeks to link "social" and "ecological" environmental concerns, while at the same time preventing de facto racism, and classism. This makes it particularly adequate for the construction of labor-environmental alliances.[3]
- Ecology movement could involve the Gaia Theory, as well as Value of Earth and other interactions between humans, science, and responsibility.
- Deep Ecology is an ideological spinoff of the ecology movement that views the diversity and integrity of the planetary ecosystem, in and for itself, as its primary value.
- Bright green environmentalism is a currently popular sub-movement, which emphasizes the idea that through technology, good design and more thoughtful use of energy and resources, people can live responsible, sustainable lives while enjoying prosperity.

Environmental law and theory

Property rights

Many environmental lawsuits question the legal rights of property owners, and whether the general public has a right to intervene with detrimental practices occurring on someone else's land. Environmental law organizations exist all across the world, such as the Environmental Law and Policy Center in the midwestern United States.

Citizens' rights

One of the earliest lawsuits to establish that citizens may sue for environmental and aesthetic harms was Scenic Hudson Preservation Conference v. Federal Power Commission, decided in 1965 by the Second Circuit Court of Appeals. The case helped halt the construction of a power plant on Storm King Mountain in New York State. See also United States environmental law and David Sive, an attorney who was involved in the case.

Nature's rights

Christopher D. Stone's 1972 essay, "Should trees have standing?" addressed the question of whether natural objects themselves should have legal rights. In the essay, Stone suggests that his argument is valid because many current rights-holders (women, children) were once seen as objects.

Environmental reactivism

Numerous criticisms and ethical ambiguities have led to growing concerns about technology, including the use of potentially-harmful pesticides, water additives like fluoride, and the extremely dangerous ethanol-processing plants.

NIMBY syndrome refers to public outcry caused by knee-jerk reaction to an unwillingness to be exposed to even necessary developments. Some serious biologists and → ecologists created the scientific ecology movement which would not confuse empirical data with visions of a desirable future world.

Modern environmentalism

Today, the sciences of → ecology and environmental science, rather than any aesthetic goals, provide the basis of unity to most serious environmentalists. As more information is gathered in scientific fields, more scientific issues like biodiversity, as opposed to mere aesthetics, are a concern. Conservation biology is rapidly-developing field. Environmentalism now has proponents in business: new ventures such as those to reuse and recycle consumer electronics and other technical equipment are gaining popularity. Computer liquidators are just one example.

In recent years, the environmental movement has increasingly focused on global warming as a top issue. As concerns about climate change moved more into the mainstream, from the connections drawn between global warming and Hurricane Katrina to Al Gore's film An Inconvenient Truth, many environmental groups refocused their efforts. In the United States, 2007 witnessed the largest grassroots environmental demonstration in years, Step It Up 2007, with rallies in over 1,400 communities and all 50 states for real global warming solutions.

Many religious organizations and individual churches now have programs and activities dedicated to environmental issues[4] The religious movement is often supported by interpretation of scriptures.[5] Most major religious groups are represented including Jewish, Islamic, Anglican, Orthodox, Evangelical, Christian and Catholic.

Radical environmentalism

Radical environmentalism emerged out of an → ecocentrism-based frustration with the co-option of mainstream environmentalism. The radical environmental movement aspires to what scholar Christopher Manes calls "a new kind of environmental activism: iconoclastic, uncompromising, discontented with traditional conservation policy, at time illegal ..." Radical environmentalism presupposes a need to reconsider Western ideas of religion and philosophy (including capitalism, patriarchy[6] and globalization)[7] sometimes through "resacralising" and reconnecting with nature.[6]

Criticisms of the environmental movement

A study reported in *The Guardian* concluded that "people who believe they have the greenest lifestyles can be seen as some of the main culprits behind global warming." The researchers found that individuals who were more environmentally conscious were more likely to take long-distance overseas flights, and that the resulting carbon emissions outweighed the savings from green lifestyles at home.[8]

See also

- Bright green environmentalism
- Carbon Neutrality
- Conservation Movement
- Earth Science
- Eco-anarchism
- Eco-socialism
- Ecofascism
- Ecoliterature
- → Ecological economics
- Ecological modernization
- → Ecology
- → Environmentalism
- Environmental organizations
- Environmental science
- → Environmental skepticism
- Free-market environmentalism
- Green anarchism
- Green movement
- Green seniors
- Green syndicalism
- Natural environment
- Political ecology
- Positive environmentalism
- Reforestation
- Radical environmentalism
- Sustainability
- Technogaianism
- Timeline of environmental events

Regional environmental movements

- Environmental movement in the United States

- Environmental movement in New Zealand
- Environmental movement in Australia

Further reading

- Paul Hawken, *Blessed Unrest*, Penguin Books Ltd, United States of America, 2007
- John McCormick, *The Global Environmental Movement*, London: John Wiley, 1995
- Ramachandra Guha *Environmentalism: A Global History*, London, Longman, 1999
- Sheldon Kamieniecki, editor, *Environmental Politics in the International Arena: Movements, Parties, Organizations, and Policy*, Albany: State University of New York Press, 1993, ISBN 0-7914-1664-X
- Philip Shabecoff, *A Fierce Green Fire: The American Environmental Movement*, Island Press; Revised Edition, 2003,ISBN 1559634375
- Paul Wapner, *Environmental Activism and World Civil Politics*, Albany: State University of New York, 1996, ISBN 0-7914-2790-0
- de Steiguer, J.E. 2006. *The Origins of Modern Environmental Thought*. The University of Arizona Press. Tucson. 246 pp.

References

[1] Theodore Roosevelt, Address to the Deep Waterway Convention Memphis, TN, October 4, 1907

[2] Most of the information in this section comes from John McCormick, *The Global Environmental Movement*, London: John Wiley, 1995.

[3] *Uniting to Win: Labor-Environmental Alliances*, by Dan Jakopovich (http://www.informaworld.com/smpp/content~db=all~content=a912901178)

[4] List of (incomplete) religious environmental organizations. (http://www.gis.net/~rwe/links.html)

[5] Biblical references related to environmentalism (http://www.creationcare.org/resources/scripture.php)

[6] Manes, Christopher, 1990. *Green Rage: Radical Environmentalism and the Unmaking of Civilization*, Boston: Little, Brown and Co.

[7] A Brief Description of Radical Environmentalism (http://shiftshapers.gnn.tv/blogs/9306/A_Brief_Description_of_Radical_Environmentalism), Jeff Luers, *4 Struggle Magazine*, 26th September 2005.

[8] David Adam, "Green idealists fail to make grade, says study," (http://www.guardian.co.uk/environment/2008/sep/24/ethicalliving.recycling) *The Guardian*, 2008-09-24

Environmentalism

Environmentalism is a broad philosophy and social movement regarding concerns for environmental → conservation and improvement of the state of the environment. Environmentalism and environmental concerns are often represented by the color green.[1]

The historic Blue Marble photograph. Environmentalism is a concern for the planet as a whole.

Environmentalism as a social movement

Environmentalism can also be defined as a → social movement that seeks to influence the political process by lobbying, activism, and education in order to protect natural resources and ecosystems. In recognition of humanity as a participant in ecosystems, the → environmental movement is centered on → ecology, health, and human rights.

An *environmentalist* is a person who may advocate the sustainable management of resources and stewardship of the natural environment through changes in public policy or individual behavior. In various ways (for example, grassroots activism and protests), environmentalists and environmental organizations seek to give the natural world a stronger voice in human affairs.

In recent years environmentalism has frequently been called, or compared to, a religion by its detractors [2]. This is usually done at the same time as pointing out what are perceived as extremist viewpoints or actions of self-identified environmentalists or groups.

History

A concern for environmental protection has recurred in diverse forms, in different parts of the world, throughout history. For example, in the Middle East, the earliest known writings concerned with environmental pollution were Arabic medical treatises written during the "Arab Agricultural Revolution", by writers such as Alkindus, Costa ben Luca, Rhazes, Ibn Al-Jazzar, al-Tamimi, al-Masihi, Avicenna, Ali ibn Ridwan, Isaac Israeli ben Solomon, Abd-el-latif, and Ibn al-Nafis. They were concerned with air contamination, water contamination, soil contamination, solid waste mishandling, and environmental assessments of certain localities.[3]

In Europe, King Edward I of England banned the burning of sea-coal by proclamation in London in 1272, after its smoke had become a problem.[4] [5] The fuel was so common in England that this earliest of names for it was acquired because it could be carted away from some shores by the wheelbarrow. Air pollution would continue to be a problem there, especially later during the industrial revolution, and extending into the recent past with the Great Smog of 1952.

See Timeline of history of environmentalism

Origins of the modern environmental movement

In Europe, it was the Industrial Revolution that gave rise to modern environmental pollution as it is generally understood today. The emergence of great factories and consumption of immense quantities of coal and other fossil fuels gave rise to unprecedented air pollution and the large volume of industrial chemical discharges added to the growing load of untreated human waste.[6] The first large-scale, modern environmental laws came in the form of the British Alkali Acts, passed in 1863, to regulate the deleterious air pollution (gaseous hydrochloric acid) given off by the Leblanc process, used to produce soda ash. Environmentalism grew out of the amenity movement, which was a reaction to industrialization, the growth of cities, and worsening air and water pollution.

In the United States, the beginnings of an environmental movement can be traced as far back as 1739, though it was not called environmentalism and was still considered conservation until the 1950's. Benjamin Franklin and other Philadelphia residents, citing "public rights," petitioned the Pennsylvania Assembly to stop waste dumping and remove tanneries from Philadelphia's commercial district. The US movement expanded in the 1800s, out of concerns for protecting the natural resources of the West, with individuals such as John Muir and Henry David Thoreau making key philosophical contributions. Thoreau was interested in peoples' relationship with nature and studied this by living close to nature in a simple life. He published his experiences in the book *Walden,* which argues that people should become intimately close with nature. Muir came to believe in nature's inherent right, especially after spending time hiking in Yosemite Valley and studying both the ecology and geology. He successfully lobbied congress to form Yosemite National Park and went on to set up the Sierra Club. The conservationist principles as well as the belief in an inherent right of nature were to become the bedrock of modern environmentalism.

In the 20th century, environmental ideas continued to grow in popularity and recognition. Efforts were starting to be made to save some wildlife, particularly the American Bison. The death of the last Passenger Pigeon as well as the endangerment of the American Bison helped to focus the minds of conservationists and popularize their concerns. In 1916 the National Park Service was founded by US President Woodrow Wilson.

In 1949, → *A Sand County Almanac* by → Aldo Leopold was published. It explained Leopold's belief that humankind should have moral respect for the environment and that it is unethical to harm it. The book is sometimes called the most influential book on conservation.

Throughout the 1950s, 1960s, 1970s and beyond, photography was used to enhance public awareness of the need for protecting land and recruiting members to environmental organizations. David Brower, Ansel Adams and Nancy Newhall created the Sierra Club Exhibit Format Series, which helped raise public environmental awareness and brought a rapidly increasing flood of new members to the Sierra Club and to the environmental movement in general. "This Is Dinosaur" edited by Wallace Stegner with photographs by Martin Litton and Philip Hyde prevented the building of dams within Dinosaur National Monument by becoming part of a new kind of activism called environmentalism that combined the conservationist ideals of Thoreau, Leopold and Muir with hard-hitting advertising, lobbying, book distribution, letter writing campaigns, and more. The powerful use of photography in addition to the written word for conservation dated back to the creation of Yosemite National Park, when photographs convinced Abraham Lincoln to preserve the beautiful glacier carved landscape for all time. The Sierra Club Exhibit Format Series galvanized public opposition to building dams in the Grand Canyon and protected many other national treasures. The Sierra Club often led a coalition of many environmental groups including the Wilderness Society and many others. After a focus on preserving wilderness in the 1950s and 1960s, the Sierra Club and other groups broadened their focus to include such issues as air and water pollution, population control, and curbing the exploitation of natural resources.

In 1962 *Silent Spring* by American biologist → Rachel Carson was published. The book cataloged the environmental impacts of the indiscriminate spraying of DDT in the US and questioned the logic of releasing large amounts of chemicals into the environment without fully understanding their effects on ecology or human health. The book suggested that DDT and other pesticides may cause cancer and that their agricultural use was a threat to wildlife, particularly birds.[7] The resulting public concern led to the creation of the United States Environmental Protection

Agency in 1970 which subsequently banned the agricultural use of DDT in the US in 1972. The limited use of DDT in disease vector control continues to this day in certain parts of the world and remains controversial. The book's legacy was to produce a far greater awareness of environmental issues and interest into how people affect the environment. With this new interest in environment came interest in problems such as air pollution and petroleum spills, and environmental interest grew. New pressure groups formed, notably Greenpeace and Friends of the Earth.

In the 1970s, the Chipko movement was formed in India; influenced by Mohandas Gandhi, they set up peaceful resistance to deforestation by literally hugging trees (leading to the term "tree huggers"). Their peaceful methods of protest and slogan "ecology is permanent economy" were very influential.

By the mid-1970s, many felt that people were on the edge of environmental catastrophe. The Back-to-the-land movement started to form and ideas of environmental ethics joined with anti-Vietnam War sentiments and other political issues. These individuals lived outside normal society and started to take on some of the more radical environmental theories such as → deep ecology. Around this time more mainstream environmentalism was starting to show force with the signing of the Endangered Species Act in 1973 and the formation of CITES in 1975.

In 1979, James Lovelock, a former NASA scientist, published *Gaia: A new look at life on Earth*, which put forth the Gaia Hypothesis; it proposes that life on Earth can be understood as a single organism. This became an important part of the Deep Green ideology. Throughout the rest of the history of environmentalism there has been debate and argument between more radical followers of this Deep Green ideology and more mainstream environmentalists.

Environmentalism has also changed to deal with new issues such as global warming and genetic engineering.

Environmental movement

The → environmental movement (a term that sometimes includes the → conservation and green movements) is a diverse scientific, social, and political movement. In general terms, environmentalists advocate the sustainable management of resources, and the protection (and restoration, when necessary) of the natural environment through changes in public policy and individual behavior. In its recognition of humanity as a participant in ecosystems, the movement is centered around → ecology, health, and human rights. Additionally, throughout history, the movement has been incorporated into religion. The movement is represented by a range of organizations, but has a younger demographic than is common in other social movements (see green seniors). Because of its large membership, varying and strong beliefs, the movement is not entirely united. Some argue that an environmental ethic of at least some sort is so urgently needed in all quarters and that the broader the movement is the better. Conversely, others assert that disunity can be a weakness in the face of strong opposition from unsympathetic political and industrial forces.

Free market environmentalism

Free market environmentalism is a theory that argues that the free market, property rights, and tort law provide the best tools to preserve the health and sustainability of the environment. It considers environmental stewardship to be natural, as well as the expulsion of polluters and other aggressors through individual and class action.

Preservation and conservation

Environmental preservation in the United States is viewed as the setting aside of natural resources to prevent damage caused by contact with humans or by certain human activities, such as logging, mining, hunting, and fishing, only to replace them with new human activities such as tourism and recreation.[8] Regulations and laws may be enacted for the preservation of natural resources.

Environmental organizations and conferences

Environmental organizations can be global, regional, national or local; they can be government-run or private (NGO). Despite a tendency to see environmentalism as an American or Western-centered pursuit, almost every country has its share of environmental activism. Moreover, groups dedicated to community development and social justice may also attend to environmental concerns.

There are some volunteer organizations. For example Ecoworld, which is about the environment and is based in team work and volunteer work. Some US environmental organizations, among them the Natural Resources Defense Council and the Environmental Defense Fund, specialize in bringing lawsuits (a tactic seen as particularly useful in that country). Other groups, such as the US-based National Wildlife Federation, the Nature Conservancy, and The Wilderness Society, and global groups like the World Wide Fund for Nature and Friends of the Earth, disseminate information, participate in public hearings, lobby, stage demonstrations, and may purchase land for preservation. Smaller groups, including Wildlife Conservation International, conduct research on endangered species and ecosystems. More radical organizations, such as Greenpeace, Earth First!, and the Earth Liberation Front, have more directly opposed actions they regard as environmentally harmful. While Greenpeace is devoted to nonviolent confrontation as a means of bearing witness to environmental wrongs and bringing issues into the public realm for debate, the underground Earth Liberation Front engages in the clandestine destruction of property, the release of caged or penned animals, and other criminal acts. Such tactics are regarded as unusual within the movement, however.

On an international level, concern for the environment was the subject of a UN conference in Stockholm in 1972, attended by 114 nations. Out of this meeting developed UNEP (United Nations Environment Programme) and the follow-up United Nations Conference on Environment and Development in 1992. Other international organizations in support of environmental policies development include the Commission for Environmental Cooperation (NAFTA), the European Environment Agency (EEA), and the Intergovernmental Panel on Climate Change (IPCC).

See also

- Conservation ethic
- → Conservation movement
- Ecology movement
- → Environmental movement
- Environmentalism in music
- Environmentalism in film and television
- Free-market environmentalism
- List of environment topics
- Positive environmentalism
- Stewardship (theology)

Further reading

- Hall, Jeremiah. "History Of The Environmental Movement [9]". Retrieved 2006-11-25.
- Kovarik, William. "Environmental History Timeline [10]". Retrieved 2006-11-25.
- Martell, Luke. "Ecology and Society: An Introduction [4]". Polity Press, 1994.
- de Steiguer, J. Edward. 2006. *The Origins of Modern Environmental Thought*. The University of Arizona Press. Tucson. 246 pp.
- John McCormick. 1995. The Global Environmental Movement. John Wiley. London. 312 pp.
- Marco Verweij and Michael Thompson (eds), 2006, *Clumsy solutions for a complex world: Governance, politics and plural perceptions*, Basingstoke: Palgrave Macmillan.

- World Bank, 2003, "Sustainable Development in a Dynamic World: Transforming Institutions, Growth, and Quality of Life" [11], World Development Report 2003, The World Bank for Reconstruction and Development and Oxford University Press.

External links

- Environment [12] at the Open Directory Project

References

[1] Cat Lincoln (Spring 2009). " Light, Dark and Bright Green Environmentalism (http://www.greendaily.com/2009/04/23/ light-dark-and-bright-green-environmentalism/)". Green Daily. . Retrieved 2009-11-02.

[2] http://www.lewrockwell.com/orig8/ostrowski-john1.html

[3] Gari, L. (November 2002). " Arabic Treatises on Environmental Pollution up to the End of the Thirteenth Century (http://www. ingentaconnect.com/content/whp/eh/2002/00000008/00000004/art00005)". *Environment and History* (White Horse Press) **8** (4): 475–488. doi: 10.3197/096734002129342747 (http://dx.doi.org/10.3197/096734002129342747). . Retrieved 2008-12-02. "This paper is limited to the works that deal with environmental pollution as a cause of such illnesses. They cover subjects like air and water contamination, solid waste mishandling and environmental assessments of certain localities. The treatises reviewed are those written by (1) al-Kindi, (2) Qusta b. Luqa, (3) al-Razi, (4) Ibn al-Jazzar, (5) al-Tamimi, (6) Abu Sahl al-Masihi, (7) Ibn Sina, (8) Ali b. Ridwan, (9) Ibn Jumay', (10) Ya'qub al-Isra'ili, (11) Abdullatif al-Baghdadi, (12) Ibn al-Quff and (13) Ibn al-Nafis.".

[4] David Urbinato (Summer 1994). " London's Historic 'Pea-Soupers' (http://www.epa.gov/history/topics/perspect/london.htm)". United States Environmental Protection Agency. . Retrieved 2006-08-02.

[5] " Deadly Smog (http://www.pbs.org/now/science/smog.html)". PBS. 2003-01-17. . Retrieved 2006-08-02.

[6] Fleming, James R.; Bethany R. Knorr. " History of the Clean Air Act (http://www.ametsoc.org/sloan/cleanair/)". American Meteorological Society. . Retrieved 2006-02-14.

[7] Carson, Rachel (1962). *Silent Spring*. Boston: Houghton Mifflin.

[8] Cunningham, William P.; et al. (1998). *Environmental encyclopedia*. Gale Research. ISBN 081039314X.

[9] http://www.mtmultipleuse.org/endangered/esahistory.htm

[10] http://www.environmentalhistory.org/

[11] http://go.worldbank.org/3I9K0DUDC0

[12] http://www.dmoz.org/Society/Issues/Environment/

Environmental skepticism

Environmental skepticism is an umbrella term that describes those that argue that particular claims put forward by → environmentalists and environmental scientists are incorrect or exaggerated, along with those who are critical of environmentalism in general. The use of the term is contested. Supporters of environmentalists argue that "skepticism" implies an open-minded attitude to empirical evidence and that their opponents are in fact advocates for predetermined positions reflecting ideological commitments or financial interests.[1] Critics of environmental skepticism frequently use more pejorative terms such as denialism.

Examples of Skeptic Thinking

The popularity of the term was enhanced by Bjørn Lomborg's book *The Skeptical Environmentalist.*[2] Lomborg approached environmental claims from a statistical and economic standpoint, and concluded that often the claims made by environmentalists were overstated. Lomborg argued, on the basis of cost benefit analysis, that few environmentalist claims warranted serious concern.

Tom Bethell's *Politically Incorrect Guide to Science* extends environmental skepticism to a more general critique of scientific consensus.[3] Bethell rejects mainstream views on evolution and global warming and supports AIDS denialism.

Frank Luntz's 2002 memo *The Environment: A Cleaner, Safer, Healthier America* to President George W. Bush emphasizes that "you need to continue to make the lack of scientific certainty a primary issue in the debate [of environmental issues]" in order to preserve the political support of voters.[4] [5] Luntz has since changed his thinking, but stands by what he said as appropriate at the time he said it.[6]

The Heartland Institute of the United States and its 'common sense environmentalism' is an example of environmental skeptic thinking. The Institute makes the distinction between 'real' environmental issues and 'imaginary' ones. It advocates that only the 'real' issues should be addressed, and done so "without trampling on other things we value, such as individual freedom and economic prosperity."[7]

Criticism

A recent study of the environmental skepticism movement found that the overwhelming majority of environmentally skeptical books published since the 1970s were either written or published by authors or institutions affiliated with conservative think tanks. The authors identified four defining themes in the movement:

1. The "denial of the seriousness of environmental problems and dismissal of scientific evidence documenting these problems"
2. The "question[ing of] the importance of environmentally protective policies"
3. The "endorse[ment] an anti-regulatory/anti-corporate liability position"
4. And the promotion of the idea that "environmentalism [is] a growing threat to social and economic progress and the 'American way of life'"

They "conclude that scepticism is a tactic of an elite-driven counter-movement designed to combat environmentalism, and that the successful use of this tactic has contributed to the weakening of US commitment to environmental protection."[8]

The mainstream news media of the United States is an example of the effectiveness of skepticism as a 'tactic'. A 2005 study reviewed and analyzed the US mass-media coverage of the environmental issue of climate change from 1988 to 2004. The authors confirm that within the journalism industry there is great emphasis on eliminating the presence of media bias. In their study they found that - due to this practice of objectivity - "Over a 15-year period, a majority (52.7%) of prestige-press articles featured balanced accounts that gave 'roughly equal attention' to the views that humans were contributing to global warming and that exclusively natural fluctuations could explain the earth's temperature increase." As a result, they observed that it is easier for people to conclude that the issue of global warming and the accompanying scientific evidence is still hotly debated.[9]

See also

- The Skeptical Environmentalist
- Bjørn Lomborg
- Penn & Teller

Bibliography

- Bethell, Tom, *The Politically Incorrect Guide to Science*, Washington, DC, Regnery Publishing, Inc., 2005 ISBN 0-89526-031-X
- Huber, Peter, *Hard Green: Saving the Environment from the Environmentalists*, New York, Basic Books, 2000 ISBN 0-465-03113-7
- Lomborg, Bjørn, *The Skeptical Environmentalist: Measuring the Real State of the World*, Cambridge & New York, Cambridge University Press, 2001 ISBN 0-521-01068-3
- Mooney, Chris, *The Republican War on Science*, Basic Books. ISBN 0465046762.
- de Steiguer, J.E. 2006. *The Origins of Modern Environmental Thought.* The University of Arizona Press. Tucson. 246 pp.

References

[1] " 'Denial lobby' turns up the heat (http://observer.guardian.co.uk/business/story/0,6903,1431306,00.html)". The Observer. . Retrieved 2008-02-07.
[2] Lomborg, Bjørn; Bjorn Lomborg (2004). *Global crises, global solutions*. Cambridge, UK: Cambridge University Press. ISBN 0-521-60614-4.
[3] Bethell, Tom (2005). *The Politically Incorrect Guide to Science*. Washington, D.C: Regnery Publishing. ISBN 0-89526-031-X.
[4] "Frank Luntz's memo at www.aspenlawschool.com" (http://www.aspenlawschool.com/books/plater_environmentallaw/updates/02.6. pdf)
[5] "Frank Luntz's memo at www.sindark.com" (http://www.sindark.com/NonBlog/Articles/LuntzResearch_environment.pdf)
[6] *"Climate chaos: Bush's climate of fear"* (http://news.bbc.co.uk/1/hi/programmes/panorama/5005994.stm). [TV Programme]. UK: BBC. 2006.
[7] "The Heartland Institute: Environment Policy and Freedom" (http://www.heartland.org/suites/environment/index.html)
[8] Jacques, P.J.; Dunlap, R.E., and Freeman, M. (June 2008). "The organisation of denial: Conservative think tanks and environmental scepticism". *Environmental Politics* 17 (3): 349–385. doi: 10.1080/09644010802055576 (http://dx.doi.org/10.1080/09644010802055576).
[9] Boykoff, M. T., Boykoff, J. M. (2007) "Climate change and journalistic norms: A case-study of US mass-media coverage." (http://www. sciencedirect.com/science?_ob=ArticleURL&_udi=B6V68-4NC5T4P-1&_user=582538&_coverDate=11/30/2007&_alid=863663502& _rdoc=3&_fmt=full&_orig=search&_cdi=5808&_sort=d&_docanchor=&view=c&_ct=5&_acct=C000029718&_version=1& _urlVersion=0&_userid=582538&md5=bbe408ce43b61e514b545aa4b21472f6#secx13), ScienceDirect Retrieved on February 07, 2009.

List of environmental philosophers

A **list of environmental philosophers**, ordered alphabetically.

This list includes living or recently deceased individuals who have published in the field of environmental ethics/philosophy (most of whom have PhDs in Philosophy, and are employed as philosophy professors), and those who are commonly regarded as precursors to the field. See below for a list of leading philosophy programs in environmental ethics/philosophy. The list should serve students, scholars, and the public interested in identifying sites of thinking on environmental questions.

→ Environmental philosophy has been impacted by individuals with a range of backgrounds, reminiscent of the approach of the natural philosophy, natural religion, and natural history traditions. Interestingly, the list of environmental philosophers includes biologists, ecologists and other scientists who have considered and written extensively about the environment and the human impact and relation to the environment.

- David Abram
- Glenn Albrecht
- Thomas Berry
- Murray Bookchin
- J. Baird Callicott, University of North Texas
- Alan Carter, University of Glasgow
- Deane Curtin, Gustavus Adolphus College
- Jan Deckers
- Martin Drenthen, Radboud University Nijmegen
- Kevin Elliott
- Paul R. Ehrlich
- Robert Figueroa, University of North Texas
- Bruce Foltz, Eckerd College
- Robert Frodeman, University of North Texas
- Steve Gardiner, University of Washington
- Harold Glasser
- Brian Goodwin
- Kenneth Goodpaster
- Roger S. Gottlieb, Worcester Polytechnic Institute
- Lori Gruen, Wesleyan University
- Pete A.Y. Gunter, University of North Texas
- James Hatley, Salisbury University
- Stephan Harding
- Eugene Hargrove, University of North Texas
- Thomas Heyd
- Dale Jamieson, New York University
- Baylor Johnson
- Robert Kirkman, Georgia Institute of Technology
- Irene Klaver, University of North Texas
- Erazim Kohak
- → Aldo Leopold[1]
- Andrew Light, Center for American Progress and George Mason University
- James Lovelock
- Alan Marshall, University of Pavol Jozef Šafárik, Slovakia

- Freya Mathews, La Trobe University
- Humberto Maturana
- Katie McShane, Colorado State University
- Carolyn Merchant
- Mary Midgley[2]
- Ben Minteer, Arizona State University
- Kathleen Dean Moore, Oregon State University
- Arne Næss
- Michael P. Nelson, Michigan State University
- Lisa Newton, Fairfield University
- Bryan G. Norton, Georgia Institute of Technology
- Max Oelschlaeger, Northern Arizona University
- David Orr, Oberlin College
- Clare Palmer, Washington University in St. Louis
- John Passmore, deceased
- Val Plumwood, Australian National University
- Christopher Preston, University of Montana
- Tom Regan, North Carolina State University
- John Rodman, Claremont College
- Ricardo Rozzi, University of North Texas
- Holmes Rolston III, Colorado State University
- Jeffrey Sachs
- James A. Sage, University of Wisconsin–Stevens Point
- Mark Sagoff, University of Maryland
- George Sessions
- Albert Schweitzer
- Kristin Shrader-Frechette, Notre Dame University
- Paul Shepard, Claremont-McKenna College
- Peter Singer, Princeton University
- Gary Snyder, University of California, Davis
- Jim Sterba, Notre Dame University
- Richard Sylvan (Routley)
- Bron Taylor, University of Florida
- Paul Taylor
- Paul Thompson, Michigan State University
- Ted Toadvine, University of Oregon
- Brian Treanor, Loyola Marymount University
- Francisco Varela
- Gary Varner, Texas A&M University
- John A. Vucetich, Michigan Technological University
- Laura Westra, University of Windsor
- Mark A. Woods, San Diego State University
- Paul M. Wood
- Michael Zimmerman, University of Colorado

See also

- Ecofeminism
- → Environmental ethics
- → Environmental philosophy
- List of philosophers
- Natural philosophy

References

[1] " The Land Ethic (http://www.aldoleopold.org/about/landethic.htm)". The Aldo Leopold Foundation. . Retrieved 2007-11-10.

[2] " Authors: Mary Midgley (http://www.giffordlectures.org/Author.asp?AuthorID=223)". Gifford Lecture Series. . Retrieved 2007-11-10.

Van Rensselaer Potter

Van Rensselaer Potter II (August 27, 1911 – September 6, 2001) was an American biochemist. He was Professor of Oncology at the McArdle Laboratory for Cancer Research at the University of Wisconsin–Madison for more than 50 years.

In 1970, Dr. Potter created the term *bioethics* to describe a new philosophy that sought to integrate biology, ecology, medicine, and human values. Bioethics is often linked to → environmental ethics and stands in sharp contrast to biomedical ethics. Because of this confusion (and appropriation of the term in medicine), Potter chose to use the term *global bioethics* in 1988.

Publications

Popular

- *Bioethics: Bridge to the Future* (Prentice-Hall, 1971)
- *Global Bioethics: Building on the Leopold Legacy* (Michigan State Univ Pr 1988) ISBN 0-87013-264-4

See also

- → Aldo Leopold

External links

- In Memoriam: McArdle Laboratory [1]
- Obituary: *Cancer Research 63, 1724, April 1, 2003* [2]

References

[1] http://mcardle.oncology.wisc.edu/faculty/bio/potter_v.html

[2] http://cancerres.aacrjournals.org/cgi/content/full/63/7/1724

Article Sources and Contributors

Environmental ethics *Source*: http://en.wikipedia.org/w/index.php?title=Environmental_ethics *Contributors*: Alan Liefting, Anthere, Anthon.Eff, Ayanoa, Badbruno, Beetstra, Bjelkeman, Bluedillygal, Bryan Derksen, Burlywood, CALR, Calibas, Ceancata, Clemifornia, Common Man, Daniel Collins, Dante Alighieri, Deeptrivia, Download, Ecomimicryproject, Ethicalpublishing, F Thome, Fasten, Former user 2, GeorgesPrat, Grucio, Guanaco, Hard Raspy Sci, Hardys, Horselover Frost, Icut4you, Iridescent, IronGargoyle, JHunterJ, JRR Trollkien, Jeepday, Jj137, Karl-Henner, Kellen`, Kozuch, Kyle Rayner, Kzollman, Leflyman, Lexor, Madhava 1947, Malleus Fatuorum, Mdd, Mlabar, Mostlyharmless, Muxxa, Mwanner, Neufast, Nopetro, Omphaloscope, Pdn, Pearle, Pengo, Peter Ellis, Prof C, Racmccaf, RexNL, Richard001, Ringbang, Salsia, SamuelAdamsOMG, Sars, Schi, SeamusSweeney, Septegram, Signalhead, Simon Fisher, Syra987, Tomisti, Vincej, Viriditas, Wavelength, Welsh, Wprlh, 101 anonymous edits

Environmental philosophy *Source*: http://en.wikipedia.org/w/index.php?title=Environmental_philosophy *Contributors*: Alan Liefting, Anarchia, Frayr, Gregbard, HisSpaceResearch, Iridescent, LilHelpa, Pahndeepah, Pengo, Tigmic, Tomisti, Vincej, 9 anonymous edits

Environmental law *Source*: http://en.wikipedia.org/w/index.php?title=Environmental_law *Contributors*: Ababich, Ado2102, AeroTraveler, Alan Liefting, Alex756, Anlace, Asm76, Barek, Beetstra, Bluxed, Bobo192, Bryan Derksen, CIELAP, CTrux23, CapitalR, Caroltroberts, Conversion script, Coopj, Ctj, D6, DASonnenfeld, Da monster under your bed, Eastlaw, Elflaw, Emote, Envirolaw, Feministo, Fluri, Forbsey, Fyyer, Graham87, Guanaco, Hard Raspy Sci, Harryzilber, Hu12, Icydesign, JNW, JRR Trollkien, JRodz15, JamesMLane, Jasonrflanders, Joeblakesley, La goutte de pluie, Lexor, MER-C, Mac, Marx01, Mbeychok, MeegsC, Metagraph, Mewmew sakura, Mintguy, Moreau1, NSWEDO, Netesq, Nilmerg, Nopetro, Npollard, Nposs, Ohnoitsjamie, Oregonlawprof, PrincessCaitlai, Redkern, Rholton, Rich Farmbrough, Richard001, Sardanaphalus, Shizane, SimonP, Simone, Sole Soul, Sortior, Stephenb, Subtle Trouble, Sunray, Superluminal23, Thunderstix, Tommyg23, Trinicheryl, Wavelength, YUL89YYZ, YURiN, Yakoo, Zer0faults, 173 anonymous edits

Environmental sociology *Source*: http://en.wikipedia.org/w/index.php?title=Environmental_sociology *Contributors*: APH, Alan Liefting, Anticipation of a New Lover's Arrival, The, Arjun01, Axionne, Bobbywhite, Clicketyclack, Cyclonedev, DASonnenfeld, Danthemankhan, Discospinster, Fishal, Goingin, Ice Czar, Iridescent, JaGa, Jeffrey Mall, LilHelpa, MichaelMBell, Mrzaius, Mspandana, Owen, Pearle, Petri Krohn, Piotrus, Planetneutral, Prof65, Richard001, SoWhy, Taishan88, Theone00, Todreamistobe, Toomanysmilies, Tous ensemble, Will, Zzuuzz, 36 anonymous edits

Ecotheology *Source*: http://en.wikipedia.org/w/index.php?title=Ecotheology *Contributors*: Aes321, Alan Liefting, Basser g, Ben Tillman, Benjiboi, Bishonen, Blue520, Bobfaricy, Calton, Cuchullain, Dakinijones, Dogaroon, Editor2020, GentlemanGhost, Gregbard, JennyRad, Jheald, Job3714, MONGO, Mgiganteus1, Nurg, Radagast83, SDC, Salix alba, Swedenborg, Tassedethe, WRK, Wavelength, Whitejay251, Wikipyrate, 25 anonymous edits

Ecological economics *Source*: http://en.wikipedia.org/w/index.php?title=Ecological_economics *Contributors*: Absinf, Altenmann, Amritasenray, Andycjp, AnnaP, AppleJuggler, Avedomni, Barak1, Behun, Berteh, Billtubbs, Biopresto, Bishonen, Black bag, BloodDoll, Bluemoose, Brighteyes0007, BruceHodge, Bthomson, Bunnyhop11, Calton, Cerealkiller13, Ceser, Charnovitz, Chris Roy, Cretog8, CryptoDerk, DASonnenfeld, DO'Neil, Dawn Bard, Dekisugi, Dingodangodongo, Dr.enh, Edindian, Ekotekk, Evb-wiki, Exiledone, F5487jin4, Fang 23, Flatbush52-1, Frankie816, Gabriel Kielland, Gary Cziko, Geronimo20, Granitethighs, Hashproduct, Hermitage, Hughcharlesparker, ImperfectlyInformed, InvictaHOG, Ixfd64, Jmalier, John D. Croft, John Pozzi, John Quiggin, Johnfos, Jorfer, Jyosna, KVDP, Kaihsu, KarlaHyde, King of Hearts, Kingturtle, Laskinmystic, Levineps, Lquilter, Lucata, M16rifle, Mac, Mailseth, Mainguy, Marcelobbribeiro, Marshman, Maurreen, Mdd, Mgw, Michael Jon Jensen, Michaello, Miriam Kennet, Miss Madeline, Moominoid, Morphh, Moulaert, Mr3641, MrRadioGuy, Mydogategodshat, Neelix, Nehrams2020, Nikai, Nopetro, Panozzaj, Pearle, Prof C, Raoulduke47, RasputinAXP, Rd232, Requestion, RichWoodward, Robbie andrew, Rodmadar, S.W. Bremer, Sade, Samsara, Sbandrews, SebastianHelm, Seraphimblade, Sholto Maud, Skipsievert, Smshaner, Snoyes, Swedenborg, The Anome, Thingg, Thomasmeeks, Timwi, Transity, Vcrs, Vegas949, Vitorsarno, Wavelength, WikieWikieWikie, WildWildBil, William Avery, Woohookitty, Xiong Chiamiov, YK Times, ידיע ר, 176 anonymous edits

Ecology *Source*: http://en.wikipedia.org/w/index.php?title=Ecology *Contributors*: (jarbarf), 131.118.95.xxx, 168..., 1B6, 200.191.188.xxx, A Softer Answer, A8UDI, ABF, AJim, APH, Abcdefghi, Abductive, Abu el mot, Acroterion, AdamRetchless, Adambro, Adrian.benko, Ahoerstemeier, Aitias, Akanemoto, Alan Liefting, Alansohn, Alex.tan, AlexPlank, Alnokta, Alpha Ralpha Boulevard, AlphaEta, AndonicO, Andre Engels, AndrewHowse, Andrewpmk, AngelOfSadness, Angr, Anonymous Dissident, Antandrus, Antaya, Anthere, Anypodetos, Aphaia, Apothecia, Apparition11, ArchStanton69, ArglebargleIV, ArielGold, Arkuat, Arnejohs, Arthur Rubin, Ascend, Ascidian, Ashertg, AshishG, Astudent, Atlant, Aua, Avnjay, BRG, BSATwinTowers, Babij, Bbolker, Bcasterline, Bcorr, Beano, Beetstra, Beherbert, Benc, Benshmen, Bentong Isles, Betterusername, Bettia, Bgs022, Big Brother 1984, Billy4862, Bluerasberry, Bobo192, Bongwarrior, Borislav, BoundaryRider, BozMo, Bremigan, BrianAsh, Bridgetn, Brooke87, Bwhity, Bws2002, CALR, CHE, CTZMSC3, Cabe6403, Cacycle, Caknuck, Caltas, Caltrop, Can't sleep, clown will eat me, CanisRufus, Capricorn42, Captain-tucker, CardinalDan, Carlosguitar, Catfish3, Cazort, Ceser, Ceyockey, Chameleon, Chaos, Chase me ladies, I'm the Cavalry, Christian List, Chriswaterguy, Chun-hian, Clayoquot, ClockworkSoul, Closedmouth, Clubmarx, Colincbn, Comet231993, Cometstyles, CommonsDelinker, Comrade jo, Conversion script, Corpx, CranfieldSAS, DV8 2XL, DVD R W, Dacyac, DarkFalls, Darrien, Darthgriz98, David Kendall, David Sneek, DavidLevinson, Dawn Bard, DeadEyeArrow, Debresser, Decoratrix, Deli nk, Delicious carbuncle, Deor, DerHexer, Diderot, DigitalNinja, Dihydromonoxide, Discospinster, Dj Capricorn, Dkeeper, Dl2000, Dmccabe, Doulos Christos, Dr Christopher Heathcote, Dracontes, Drat, Dreadstar, Dulin5, Dycedarg, Dysepsion, EPM, Earthdirt, EcoForecast, Ecolover2, Ecoplus, Ed Poor, El C, Eleassar777, Eliezg, Elijahmeeks, Eliyak, Elmerfadd, Encycl wiki 01, Epbr123, Erkan Yilmaz, Escape Orbit, Espoo, Evil saltine, Exert, Exlibris, Falcon8765, Fanghong, Febg, Fences and windows, Fieldday-sunday, Flubeca, Fnlayson, Forbsey, Gail, Gaius Cornelius, Gaurav, Gdk2005, Geniac, Germen, GhostPirate, Giftlite, Gimme danger, Glane23, Glanthor Reviol, Gnfnrf, Go for it!, Godbruther, Gogo Dodo, GraemeL, GregorB, Greudin, Gscshoyru, Guettarda, Gurch, H-ko, HaeB, HalfShadow, HamburgerRadio, Hannahmarieg, Hard Raspy Sci, Hdt83, Hdynes, Heero Kirashami, Heidipearson, Hexwings, Hhbruun, Hu12, Husond, IanCheesman, Ianf, Ignotus91987, ImperfectlyInformed, Information Habitat, Inkington, Int.leadership, Inter, Into The Fray, Iralith, Iridescent, Ironbark, Island, Itsmejudith, Ixfd64, J.delanoy, J04n, JForget, JYolkowski, Jack-A-Roe, Jackol, Jade Knight, Jannex, Javert, Javierito92, Jaxsonjo, Jbergquist, Jcw69, Jeannie kendrick, Jeff G., JeffreyGomez, JimR, Jj137, Jmeppley, Jnc, Jnothman, Joe hill, JoeBoucher, Joel7687, John D. Croft, John Nevard, John254, Joinn, Jojhutton, Jorge Stolfi, JoseJones, Joseph.Kidder, Joymmart, Joyous!, Jrbouldin, Juan M. Gonzalez, JustAGal, Justin.Johnsen, KAM, KPH2293, Kallimina, Karol Langner, Kathyedits, Kbdank71, Kenosis, Khalidkhoso, Kirachinmoku, Kirk Hilliard, KnowledgeOfSelf, Koliya1985, Kowey, Kubigula, Kukini, La goutte de pluie, Lalamickey123, Laurel Bush, Laxology, Lexor, Libertyblues, Lkinkade, Llull, Lmcelhiney, Looxix, Luk, MECU, MONGO, MPF, MPRO, Mac, MacTire02, Macy, Magnus Manske, Mailseth, Malcolm Farmer, Mani1, MarkSutton, Marshman, Martin451, MartinHarper, Mary Calm, Maser Fletcher, Mattworld, Maurreen, Mav, MaxEnt, Mayumashu, Mbaha, Mdd, Mejor Los Indios, Mendaliv, Mentifisto, Meownyon, Michael Hardy, Michal Nebyla, Mikeybabel01, Mimzy1990, Minnesota1, Miquonranger03, Mjmcb1, Mlee1224, Mnc4t, Mokus, Monosynthluv, Morrad, MrFacchin, MrTrauma, Mrshaba, Ms.kathleen, Murphylab, Mwanner, Mxn, Myanw, Mysdaao, N96, Nadinebrinson, Nadler, Nads27, Nairobiny, Nakon, Nakos2208, Naught101, Neffm, Nenohal, NeoJustin, NeoXNeo, Neutrality, Nhelm83, Nick C, Nightkey, Nikai, Nivix, NorwegianBlue, Notafly, Nov ialiste, Nrawat, Nufy8, Nuttycoconut, Od Mishehu, Oda Mari, Ohsokulio, Oikoschile, Oleg Alexandrov, Olivier, Omicronpersei8, Omnipaedista, Onco p53, Orangemarlin, Oxymoron83, Palaeovia, Paleorthid, Pavel Vozenilek, Peak, Pedro, Pengo, Pentasyllabic, Persian Poet Gal, Peter Wallack, Petri Krohn, Pharaoh of the Wizards, PhilKnight, Philip Trueman, Phlebas, Pinethicket, Pinkadelica, Pip2andahalf, Plumbago, Pneukal, Pollinator, Poor Yorick, Pro bug catcher, Prodego, ProfMilton, Quintessent, Quintote, Qxz, RHaworth, RJBurkhart, Raddant, Radiofreebc, Rajeevmass, Raven in Orbit, RayTomes, RazorICE, Razorflame, Recognizance, RedWolf, Redhookesb, Reginmund, Remi0o, Reo On, Res2216firestar, RexNL, Rfl, Rholton, Rich Farmbrough, Richard D. LeCour, Richard001, Ristcason, Rje, Roberta F., Robinh, Rodhullandemu, Roland2, Ron Ritzman, Ronhjones, Ronline, Rubicon, SEWilco, SJP, ST47, Samini123, Sander Säde, Santa Sangre, Saros136, Scarletspider, Sceptre, Scetoaux, SchfiftyThree, Scientizzle, Sciurinæ, Shaker john, Shinobaren, Shirifan, Sholto Maud, Shrumster, Shyamal, SigmaEpsilon, SiobhanHansa, Sjorford, Skier Dude, Skittleys, Skysmith, Smithfarm, Smshaner, Sniperspree911, Snowmobiler120, Snowolf, Solipsist, Sortior, South Bay, Spiderbo13, SpyMagician, Stanskis, StaticGull, Stefeyboy, Steinsky, Stephenb, Steverapaport, Stevertigo, Streetball-PLAYER, Sunray, Svetovid, THEN WHO WAS PHONE?, TUF-KAT, Taw, Tcncv, The Cat and the Owl, The Thing That Should Not Be, The Transhumanist, The Transhumanist (AWB), TheNewPhobia, Theanthrope, Thingg, Thompsma, Thue, Timir2, TimothyPilgrim, Tio andy, Titanbanger, Titoxd, Tivedshambo, Tmol42, Tombomp, Tombstone, Tonderai, Tony Sidaway, Topbanana, Tpbradbury, Tree Biting Conspiracy, Treisijs, Triona, Tslocum, Twirligig, Tylerisgod, Ubardak, UberScienceNerd, Ucanlookitup, UkPaolo, Unschool, Utcursch, UtherSRG, Uvo, Van helsing, Veinor, VerticalDrop, Vertoch, Vgy7ujm, Victorgrigas, Videoecology, Viriditas, Vsmith, Wachholder0, WadeSimMiser, Warlord Tsunami, Waynem, Wetman, Wevets, Whatiguana, Wik, Wiki alf, WikipedianMarlith, Wilke, William S. Saturn, Willking1979, Wimt, Wisden17, XJamRastafire, Xaosflux, Xy7, Yahel Guhan, Yangyang2036, Yansa, Yellowcab643, Zacharie Grossen, Zigger, Zzuuzz, فاع بوبحم, 1193 anonymous edits

Environmental geography *Source*: http://en.wikipedia.org/w/index.php?title=Environmental_geography *Contributors*: -Midorihana-, Alan Liefting, AlexD, Bsdlogical, CALR, Can't sleep, clown will eat me, Corvus cornix, DASonnenfeld, DerHexer, Dexter siu, Dr. Blofeld, El C, Epbr123, Falcon8765, Granitethighs, Grutness, Iridescent, J.delanoy, Jaknouse, JazzyMazzy, Jusdafax, Lando Calrissian, Maurreen, MeltBanana, Menchi, Ming28, Oxymoron83, Pearle, Rajeevmass, Reinyday, RichardF, Rolo08, Stephenb, Sydneyfc1, TrbleClef, Vsmith, Zundark, 84 anonymous edits

Clearcutting *Source*: http://en.wikipedia.org/w/index.php?title=Clearcutting *Contributors*: 16@r, 21655, AdjustShift, Alan Liefting, Andersmusician, Andrewpmk, AtheWeatherman, Bobianite, Bradjamesbrown, Brian Crawford, Cdcdoc, Cewvero, Closedmouth, ColtM4, Csab, D, Dabean, Diagonalfish, ESkog, Edward, Fabiform, Face, Failure.exe, Fieldday-sunday, Gomm, Hadrianheugh, IronGargoyle, J.delanoy, JR98664, James086, JamesAM, Ju9ijjhibjuyhbo, Juliancolton, Kehrbykid, Ken Thomas, Kleinzach, KnowledgeOfSelf, Kummi, Little Mountain 5, MONGO, MadcapLaugher, McSly, Michael Devore, Mifter, Minnecologies, Oxymoron83, Papppfaffe, Pharaoh of the Wizards, Philip Trueman, Pinethicket, Plumpurple, Poeloq, Possum, Quinnjaron, Raven in Orbit, Red, Rjwilmsi, Ryan shell, Sharkface217, Silverchemist, Stephenb, THEN WHO WAS PHONE?, The Anome, Thingg, TicketMan, Timdukes1990, Tjmcevoy, Tohd8BohaithuGh1, Triwbe, Vert et Noir, Wavelength, Wuhwuzdat, 197 anonymous edits

Internal combustion engine *Source*: http://en.wikipedia.org/w/index.php?title=Internal_combustion_engine *Contributors*: .45Colt, 1337 MJSU, 2over0, 30mph, 7, A Softer Answer, A8UDI, AGToth, Aabbcc334455, Aadal, Abomasnow, Acalamari, Addshore, Aecis, Aeglyss1, Ahoerstemeier, Akamad, Aksi great, Alan.ca, Alansohn, Albedo, Aldie, Ale jrb, Alemily, Allischalmers,

Altermike, Altes2009, Alureiter, Ampmouse, Andonio, Andre Engels, Andrewa, Andrewcool, Andromeda451, Andy Christ, Anonymous editor, Antandrus, Anthony Appleyard, Arakunem, Ardonik, Arminius, Arpingstone, Arthur Rubin, Arvindj227, AtholM, Attilios, Avriette, BD2412, BMT, Backslash Forwardslash, Balthazarduju, Bayerischermann, Beefcalf, Beland, BenFrantzDale, Bibinboy, Bigdumbdinosaur, BilCat, Billymac00, BitterMan, BlaiseFEgan, Blake, Blowdart, Bnkausik, Bobblewik, Bonaparte, Bozdog, Brandymae, Brianjd, Brion VIBBER, BritishWatcher, Bryan Duggan, Bunthorne, CDN99, CIreland, CME94, CSvBibra, Calabraxthis, Calvin 1998, Can't sleep, clown will eat me, CanadianLinuxUser, Canistabbats, Canzo, Capricorn42, Captain panda, CardinalDan, Catgut, Cbh, Cbraga, Cedars, Chagai, Chairboy, Chancemill, Chantoke, Chovain, Chraja2004, Chriom, Chris the speller, Chrislonghurst, Christian15, Christopher Parham, Citicat, Clarkbhm, Clivedog, Closedmouth, CodaBaboda, ColinPGarner, Cometstyles, Compuneo, Conversion script, Cookiecaper, Cool3, CosmicFalcon, Cpl Syx, Crateengine, Crazybearr, CyclePat, Cyrillic, D0762, D0li0, DAMurphy, DCEdwards1966, DVD R W, Daarznieks, Davewild, David R. Ingham, Dcb1995, Delirium, DerHexer, Dialectric, Diceman, Dieselcruiserhead, DigitalC, Dillinger s, DinosaursLoveExistence, Discospinster, DmitryKo, Dolphin51, Dominiklenne, Dpm64, Dqeswn, Drbdavid, Duk, Dv82matt, E Wing, EagleFan, Ear59pf, Edgar181, Edgarde, Edison, Edward, EdwardG, Eirik, Ekiledal, Elassint, Element16, Eleschinski2000, Epbr123, Epicidiot, Equendil, Escape Artist Swyer, Espoo, Euchiasmus, Euweng Chan, Everyking, Excaliburo, Excirial, Fahadsadah, Femto, Ferkelparade, FiggyBee, Fingers-of-Pyrex, Firebug, Fletcher, Flewis, Fred Bauder, FredStrauss, Fredrik, FreplySpang, Fritzpoll, FutureUApilot, Fyre4ce, G-rad, GCarty, GRAHAMUK, GTBacchus, Gabeg646, Gaius Cornelius, Ganesh.jkumar, Gblandst, Gege1234, Gene Nygaard, Geologyguy, Ggisalegend, Giftlite, Gisguy, Glenn, Gm3139, Gogo Dodo, GraemeL, GraemeLeggett, Grahamsteven, Gralo, Greglocock, Gregorof, Grn2010, GrooveDog, Grubber, Gunnar Larsson, Gurchzilla, Guyonthesubway, Gzuckier, Hadal, Hailuatyan, Hamiltondaniel, Hans-J. Wendt, Harald Hansen, Hartford57, Hbackman, Heavenly Spawn, HenryLi, Heron, Hinrik, Historikos, Hondarox715, Htra0497, Hu, Hu12, Husond, Hut 8.5, HybridBoy, Hydrox, I dream of horses, IDR-Technology, IRP, Ian01, Iburnh2o, Icseaturtles, Ildefonso Yonto, Ingolfson, Interiot, InternetHero, Inwind, Inyourface1551, Isis4563, Isomorphic, Ixfd64, J.delanoy, J2thawiki, JForget, JLaTondre, JSpung, JackLumber, Jackol, Jaganath, Jagged 85, Jake Wartenberg, James086, JamesBurns, Jasminakocan, Jbostin, Jcomp489, Jdforrester, Jennavecia, Jimfbleak, Jlenthe, Johanfo, John, John254, Johnleemk, Jon513, Jrockley, Juliancolton, Jusdafax, KC., KGV, KLLvr283, Kalper, Karn, Karpada, Karsini, Kazrak, Kdakin3420, Keegan, Kelisi, Kimse, Kingpin13, Kjkolb, KnowledgeOfSelf, Kowey, Kram-bc, Kruckenberg.1, Kshpitsa, Kungfuadam, Kwsn, LMB, LOL, Laddiebuck, Lauciusa, Lee J Haywood, Leevanjackson, Lensovet, Leonard G., Lessthanthree, Lid, Life of Riley, Liftarn, Lightmouse, Lights, Lincspoacher, Little Mountain 5, Loren.wilton, Loren36, Lowellian, Luckyleafus, Lumbercutter, Lumos3, Lupinoid, MGTom, MJCdetroit, Mac, Magnus Manske, Mailer diablo, Makemi, Malbeare, MalcolmMcDonald, Malcolma, Mark.murphy, Marky 1976, Martin451, Matt Gies, Maury Markowitz, Mav, Maximaximax, Maximus Rex, Mboverload, Mentality, Mentifisto, Meske, Michael Devore, Michael Hardy, MightyWarrior, Mikelantis, Mishuletz, Mjquinn id, Mmxx, Modulatum, Mojodaddy, Monkeymanman, Mormegil, Morther, Morven, Motorhead, Mr Adequate, Mrsillybilly, Muchosucko, Muhends, Mushroomman007, Muskisold, Mwarren us, NRen2k5, Nacor, Nakon, Nancy, Nestrus2004, Neverhead, Newfoundlanddog, NickW557, Nigholith, Nihiltres, Nikhil Sanjay Bapat, Nima rasouli, Nopetro, Nrbelex, Nseidm1, NuclearWarfare, Nuno Tavares, OMCV, Oberst, Obnoxin, Oldironnut, Omegatron, Orangutan, Owen, Oxymoron83, PDH, PFHLai, Paddy8s, Panoulis85, Patrick, Paul W, Pdelong, Per Honor et Gloria, Peter Horn, Peterlin, PhilKnight, Philip Trueman, Piano non troppo, Pjbflynn, PlatinumX, Poeloq, Pol098, Pollinator, Prentice.Wood, Proteus, Psy guy, Pyrotec, Q43, Quamaretto, Quendus, R'n'B, RB972, RPellessier, RTC, RaCha'ar, Radiojon, Raul654, Rbelli, Rdsmith4, RealGrouchy, RedJ 17, RedWolf, Reddi, Reedy, Retired username, RexNL, Rhombus, Rich Farmbrough, Richi, Rlandmann, Roastytoast, Rob Skinner, Robertvan1, Romanm, Ronaldomundo, Roscoe x, RottweilerCS, RoyBoy, Rror, Rsm99833, Rsteif, S3000, Sabudis, Sadi Carnot, Saintrain, Salasks, Salvor, Savidan, Scarian, Sceptre, SchfiftyThree, SchuminWeb, Schweiwikist, ScotchMB, Scythe33, Sdsds, Seano1, Seb az86556, Selkem, Shanes, Shawn in Montreal, Sheeson, Shoejar, Shreditor, Sietse Snel, Silverxxx, SineWave, SiobhanHansa, Sitenl, SkyWalker, Slakr, Slambo, Slashme, Smalljim, SmokeySteve, Smt52, Sno2, Sonett72, Sordomudo11, Soulkeeper, Sowilo, Spike Wilbury, SpuriousQ, Srschu273, Staffwaterboy, SteinbDJ, Stephan Leeds, SteveBaker, Stevedix, Stevenakis, Superbeecat, Superblink41, Surfturf, Systemf, TCBTCBTCB, THEN WHO WAS PHONE?, Tabletop, Tainter, Tarquin, Taw, Taxman, Tblackma222, Tea with toast, Tedder, Tellerman, Teratornis, The High Fin Sperm Whale, The Parting Glass, The Thing That Should Not Be, Thedjatclubrock, Theuion, Thingg, Thomas27, Tide rolls, Titoxd, Toast73, Toeguy, Tokino, TomRawlinson, Tomandlu, Tombstone, Tomruen, Topory, Towel401, Trevor MacInnis, Tricky Victoria, Triwbe, Trusilver, Tt45235, Turgan, Twang, Twested, Tynetrekker, Tysto, U, UB65, Ugbrad, Ukexpat, Ulric1313, Uncle Dick, Unfocused, Unvjake, UrbanTerrorist, V8Cougar, Van helsing, Vatsan34, Vcelloho, Versus22, Victuallers, Vifteovn, Vikingstad, Vina, Visor57, Vssun, WadeSimMiser, Wafulz, Waldorf, Wambo, Wapcaplet, Warmblood, Wayward, Wdl1961, Werdna, Wi-king, Wiki alf, Wikimike, WikipedianProlific, Wikisarah, Willtron, Wireless Keyboard, Wmahan, Wolfkeeper, Wrp103, Ww2censor, Wysprgr2005, Xylix, Yadyn, Yamamoto Ichiro, Zain Ebrahim111, Zeromaru, Zfr, Zhangzhe0101, Ziusudra, Zntrip, Zootsuits, Zzuuzz, Александър, Սահակ, 1475 anonymous edits

Extinction *Source*: http://en.wikipedia.org/w/index.php?title=Extinction *Contributors*: Adam78, Adamfinmo, Aero Flame, Aesopos, AgentPeppermint, Ahoerstemeier, Alan Liefting, Alansohn, Amaltheus, Amcbride, Animal-addict, Anlace, Anonymous Dissident, Antandrus, Antarticstargate, Anthony Kern, Antonio Lopez, Apollyon48, Archos, Ascánder, Ashmoo, Asidemes, Astudent, Atulsnischal, Aziz1005, Baconfish, Balbers, Ballista, Bantman, Barrylb, Bender235, Bendzh, Betacommand, Betterusername, Bettia, Big Bird, Bigmak, Bill37212, BinaryTed, Blathnaid, Bob98133, BobCatBobDog, Bobo192, Bochkov, Bookinhand, Borgx, Bradjamesbrown, Brian A Schmidt, BrianGV, Broomballcory, Brunnock, Bryan Derksen, Buffon4life, Burlywood, Burntsauce, Calaschysm, Caliix3, Calvin 1998, CamdogXIII, Capricorn42, Cavrdg, ChadThomson, CharlotteWebb, Chasbegley, Chaustin817, Chris huh, Chris the speller, Christian415, Christy747, Ckatz, Closedmouth, Cntreras, Complex (de), Cool Hand Luke, Crisneda2000, Cyberied, CzarB, D99figge, DGG, DJ Clayworth, Dana boomer, DancingMan, Daniel Brockman, DanielCD, DanielEng, Danny, DannyDaWriter, Dante Alighieri, Dantheman531, Davewild, David.Mestel, Dbergan, Dekisugi, Dendodge, DianaGaleM, Discospinster, Discworldian, Doc Tropics, DocWatson42, Dookieturd, Dot Bitch, Dreadstar, Driftwoodzebulin, Dunee, Dusti, Dysmorodrepanis, Ed Fitzgerald, Ed g2s, El C, Elvey, Emerson7, Emmiecmoore, Endangered11, Epbr123, Eras-mus, Escape Orbit, EventHorizon, Everyking, Excirial, FCStaehle, Falconleaf, Flewis, Fnarf999, ForestDim, Fornadan, Fr564, Fredrik, Funandtrvl, Fusionmix, Gail, Garethfoot, Gary King, Gbaor, Gentgeen, Gibbo25, Gogo Dodo, GraemeL, Green Giant, Greenbling07, GregAsche, Guanaco, Gurch, Gwilcox, H2g2bob, HRIN, Halmstad, HeatherTudor, Hester13, Hibernian, Hirub, HistoryBA, Hnsampat, Hq3473, Hypercane, Iant, Ibagli, IceUnshattered, Int3gr4te, Inukai, Ioexplore, Irishguy, J.delanoy, JCSantos, JRR Trollkien, JT22AC, Jackfork, JanetSmile, Jiy, Jkelly, Jkintree, Jnp2109, Joelr31, Johann Wolfgang, John Callender, John Hill, John254, Jonathansfox, Jonnyboy8807, Joseph Solis in Australia, Just H, Kalamkaar, Keegscee, Khoikhoi, Kku, Kollision, Kosigrim, Kuru, La goutte de pluie, Law, Lesqual, Lexor, Liface, Liftarn, Lightmouse, Lilac Soul, Limulus, Linkminer, Lir, Livingtree, LizardJr8, Lololololl12345678, Lord Hawk, Lowellian, LtNOWIS, MPerel, Madhero88, Mailseth, Malix, Malkinann, Manaqueen, Mandarax, Martin451, Martpol, MaryBowser, Mathwizard1232, Mav, Mcapplbee, Mejor Los Indios, Mekeretrig, Melly42, Meridius, Merurider, Mgiganteus1, Michael Hardy, Mike Rosoft, Mirv, Mmcknight4, N-k, Nacen, Neale Monks, Nihiltres, Nk.sheridan, Ohnoitsjamie, Olivier, Omicronpersei8, Orangemarlin, Overkill399, Oxcorp, Parakalo, Patrick, PenQuince, Pengo, Persian Poet Gal, Phatom87, Philip Trueman, Plumbago, Pmaas, Pmlineditor, Politoed666, Pollinator, Porqin, PrimeCupEevee, PuzzleMeister, QmunkE, Qwasty, R. fiend, Raine-07, RandomP, Randwicked, RapidR, Raz1el, Realm of Shadows, Reinoutr, RexNL, Rgoodermote, Rich Farmbrough, Richard001, Rickjpelleg, Rjwilmsi, Robinh, RoboAction, RoyBoy, Rursus, Ruud Koot, SEWilco, Sam Li, Sandahl, Sarahbriarmoss, Sat liverpool, Sceptre, SchuminWeb, Seans Potato Business, Sekolov, Siim, Siroxo, Skarebo, Sluzzelin, Smalljim, Smartse, Smihael, Smitinscience, Snigbrook, Snillet, Snowmanradio, Somagu, Spellcast, Spencerw, Stephenb, Stereotek, SteveBaker, Storm Rider, Strenshon, Streona, Stumps, Stw, SuperJumbo, Svick, Syp, Tannin, Tennis Dynamite, Terrifictriffid, The Dan, The Next Biggish Thing, The loominator, TheScurvyEye, Theinsomniac4life, Thinboy00, Thue, Thunder8, Tiddly Tom, Toddcs, Toddst1, Tognopop, Toh, Tom4216697, Tpbradbury, Trevor MacInnis, Treyt021, Triona, Tsemii, UbuntuUser2909, Vague Rant, Vicenarian, Vicki Rosenzweig, Voyevoda, Vsmith, Waggers, Wikiscient, Wildhartlivie, Wilson44691, Wimt, Winchelsea, Winkwink, Wisdom89, Woohookitty, Wragge, YAG490, Zink Dawg, 638 anonymous edits

Rachel Carson *Source*: http://en.wikipedia.org/w/index.php?title=Rachel_Carson *Contributors*: 21655, 2D, 83d40m, 999bob999, A.V., A1000, AEMoreira042281, Abaumination, Acalamari, Acroterion, Addshore, Agil, AgnosticPreachersKid, Alansohn, Albanaco, AlexWaelde, Alexius08, AlexiusHoratius, Alison, Alison22, All Is One, Allstarecho, Alphachimp, Amydmoore, Anaxial, AndonicO, Andycjp, Angmering, Antandrus, Apparition11, Ari Rahikkala, Armymp0165, Ashlandchemist, Aspire3623WXCi, Astuishin, Attackerkill, AuburnPilot, Awadewit, Awesome37, Bamadawg, Barfooz, Barneca, Bcsurvivor, Beginning, Belinrahs, Bhadani, Binary TSO, Blainster, Blanchardb, Bluemoose, Bmwhite96, Bobo192, Bogey97, Booksmedia, Boothy443, Bowlhover, Brblweb, BrianGV, Brianga, Brighterorange, Brookiegrl1996, C mon, C'est moi, CPacker, CWii, Caeruleancentaur, Calmer Waters, Caltas, Can't sleep, clown will eat me, Canderson7, Capricorn42, Caracaskid, Cassantime, Casull, Catgut, CatherineMunro, Cbrown1023, Ceyockey, Cflann01, Cgingold, Chicheley, Chris huh, ChrisGriswold, Christian Historybuff, Chun-hian, Ciaccona, Circeus, Clarkbhm, Cogizu, Cronos1, Cst17, Cuppysfriend, CurtisJohnson, Cyclopaedia, D Monack, D6, Danny, Danspalding, Darth Panda, Darwinek, Dave souza, Davidwr, Dblandford, DeadEyeArrow, Deltabeignet, Dfiwfj450, Dh100, Dimadick, Dj Capricorn, Doncram, Dondiasco, Dorvaq, Doug, Dougweller, Doulos Christos, Download, Dr Dec, DragonflySixtyseven, Dsp13, Dwalls, Dysepsion, ERcheck, Ecoleetage, Ed Poor, Eeekster, Either way, Elakhna, Elefuntboy, EllGeeOhh, Elsaandglow101, Emc2, Epbr123, Equendil, Erkan Yilmaz, Escape Orbit, Everyking, Ewen, Excirial, FWBOarticle, Fanghong, Faramir Fett, FemChatham, Fireice, Fishcomm, FisherQueen, Flcelloguy, Flockmeal, FlyingToaster, Freakofnurture, Fretlessb, FromFoamsToWaves, Fullobeans, Fyyer, Gh, Gimme danger, Gloriamarie, Golbez, Good Olfactory, Graham, Greentopia, GrooveDog, Ground Zero, Gurch, Gwernol, Gzuckier, HOUZI, Hadal, Hall Monitor, Hard Raspy Sci, Harel, Harland1, Henryhartley, Hertz1888, HexaChord, Hgilbert, Hjal, Homagetocatalonia, Howardjp, Hulagurl2008, Human.v2.0, Huntress600, Hydrogen Iodide, II MusLiM HyBRiD II, IRP, Iamthebestever9, Ibagli, Icestorm815, Icewedge, Igoldste, International1, Into The Fray, Ipwn34563, Isis, Ixfd64, J.delanoy, JForget, JNW, Jacobolus, James 6030, Janisbeth, JayHenry, JayJasper, Jckrull, Jcw69, Jeanenawhitney, Jedudedek, Jeendan, Jeff G., Jeffq, JessicaisRed, Jgspeck, Jhalpern, JimVC3, Jj137, Jjlorenzrn1, JodyB, Joe Wreschnig, John Quiggin, John254, Johnfos, Jojhutton, Jojit fb, Jomamma0828, Jonajule, Joseph Solis in Australia, Josephabradshaw, Juliancolton, Just Jim Dandy, Jwissick, Kaldari, Katerenka, Kb9wte, Keeper of Maps, Keilana, Kevinkillhill, King Toadsworth, King of Hearts, King1234567891, Kingturtle, Kiss of death01, Kkoolpatz, KnowledgeOfSelf, Krellis, Kricxjo, Kristof vt, Kuru, Lajsikonik, LaszloWalrus, LeaveSleaves, Leepaxton, Leifern, LeilaniLad, LenBudney, Lightmouse, LilLexiBaby, Lisatwo, Littera9, Lockley, Lord Vader, Lquilter, Ludde23, Luis Fernández García, Luk, Lycurgus, Lyssarox15, MBisanz, Madhero88, MajorRogers, Malatesta, Malik Shabazz, Mappychris, MareMares, Martian.knight, Martin451, MastCell, Mattbrundage, Mattisse, Maxim, Mdd, Mentifisto, Mercunis, Metta Bubble, Michael A. White, Michael Devore, MichaelSH, Mikeo, Mild Bill Hiccup, Mimihitam, Mistermango, Mjc209, Mmburt1, Mmcannis, Mnwolf, Moeron, Moni3, Moomoopoo, Mr. Know-It-All, Mr. Lefty, NW's Public Sock, Nakon, Nathanthompson, NawlinWiki, Neurolysis, Neutrality, Nichlok, Nick Carraway, Niteowlneils, Nsaa, NuclearWarfare, OPMaster, Oda Mari, Odikuas, Oldegrrl1958, OnTheGas, Outriggr, Oxymoron83, PDH, Parhamr, Pascal.Tesson, PatGallacher, Paul A, Paul Benjamin Austin, Pde, Peripatetic, Person man345, Petri Krohn, Phgao, PhiLiP, Philip Trueman, Physchim62, Pigsonthewing, Pilotguy, Pinkville, Pnosker, Poindexter Propellerhead, Pollinator, Poopoop5, Prashanthns, Prodego, Proudroy, Proyster, Pstuart84, Qthorse922, Qxz, RJaguar3, Rachellouise, Radagast3, Radon210, Ragesoss, Raime, RandomOrca2, RandomP, Raonisousa, Rich Farmbrough, Richard001, RichardF, Robert Merkel, RobertG, Rocket000, Ronhjones, Ronnotel, Rudowsky, Rufous-crowned Sparrow, Runningonbrains, Rursus, Ryryrules100, SURIV, Sam Korn, Sampodog, Sams56, Samuraikeith, Sandahl, Sandman 445sss, Schmidtr1, Schzmo, Senator Palpatine, Sethant, Shalom Yechiel, Shanes, ShelfSkewed, Shirulashem, Shoessss, SimonP, Sintaku, Sir soham, Slinga, Smallweed, Snowolf, Solarusdude, Solon.KR, Somebodysomething, Spiff, Sputnikcccp, StaticGull, Stemonitis, Stephen C. Carlson, Stephenb, SteveMcCluskey, Steven Weston, StradivariusTV, Superadvancepet, SusanLesch, Syalpranav, Symon, TJ aka Teej, Tabletop, Tannin, Tedder, Tempodivalse, The Evil Spartan, The Wordsmith, TheKMan, Thingg, Threeafterthree, Tide rolls, TimLambert, Tiptoety, Tony1, Tpbradbury, Treisijs, Trevor MacInnis, TylerScotWilliams, Uncle Dick, Uriel8, Uruiamme, VAcharon, Varano, Viriditas, Vision Thing, Waggers, WatchAndObserve, Wavelength, Weefolk29, Wetman, Weyes, Wfeidt, Wiki alf, William Avery, William M. Connolley, Willking1979, Wombatcat, Wompa99, Wonderactivist, Wqlister, Xiahou, Y2014jaceld, Yamamoto Ichiro, Yilloslime, Yuckfoo, Александр Сигачёв, №Ojuednfjewn№№, 1119 anonymous edits

Earth Day *Source*: http://en.wikipedia.org/w/index.php?title=Earth_Day *Contributors*: 09lamin, 16@r, 2ScoopRice, ACupOfCoffee, AI, Abigailb9, Abrech, Acalamari, Adam1729, Addshore, AdnanSa, Ahoerstemeier, Aitias, Alan Liefting, Alansohn, Aldaron, Algebraist, All Is One, Allstarecho, Ams80, AndonicO, AndreasJS, Anonip, Antonio Lopez, Ary29, Asanchez1572, Ashburn145, Ashley Pomeroy, Atif.t2, Attoman, Auric, Autumnalequinox, Ayanoa, Ayt999, Bantosh, Bargholz, BaseballDetective, Bbpen, Beaver, Bedford, BeeArkKey, Benleehoyin, Billwilson5060, BinaryTed, Birdman1, Bobamnertiopsis, Bongwarrior, Boothy443, Borgx, Bruce wasserman, Bsimmons666, Burkinaboy, Bwsmoney, Béka, CUSENZA Mario, CWY2190, Cacophony, Calamitybrook, Caltas, Caltrop, Canderson7, Canterbury Tail, Capsela, CardinalDan, Cashewbrick, Casito, Ceran, Cfsenel, Cgingold, Chanakal, Chelydra, Christian List, Chun-hian, Ciaccona, Citizenkoenig, Civilizededucation, Clarifythis, Cliff smith, Cmdrjameson, Cobaltbluetony, Coffee, Cometstyles, CoolRod41, Cquan, Crackerjack, CrazyChemGuy, Crunch, CryptoDerk, Cureden, Cwlq, D-Katana, DVD R W, Dagen, Daniel Earwicker, Daniel5127, David Kernow, David Underdown, Davidcannon, Dawn Bard, Deor, DerHexer, Dickens10, Dkreisst, Dlab95, Doctorfluffy, Dogposter, Dom&fallon, Donald Hosek, Doniago, Dorftrottel, DougsTech, Download, DreamGuy, Dreitmen, Drilnoth, Dust Filter, Dwalls, Dycedarg, Dylan38, Dysepsion, E Wing, E0steven, EKMichigan, Earth, Earthdaystinks, Earthdirt, Earthlyideas, Eddy1992dog, Edison, Edokter, Ejnogarb, Emeraldcityserendipity, Emijrp, Emurphy42, Endangeredplanet, Energizer07, Enviropearson, Epbr123, Equine812, F, FJPB, Fag&jesse, Fagles, Fanatix, Fbv65edel, Feydey, Flyguy649, Fonzy, Foodgreen, Forest26M, Freakmighty, FrenchyFries, Funnyhat, G4gamer05, GRIFFEY366, Gamaliel, Garden, Garrisonroo, Garzo, Gelatart, Gemmi, Geologyguy, Georgeryp, Golbez, Goldom, Graham Jones, Graham87, Gregreyj, Griffinofwales, Grumpyyoungman01, Grundle2600, Gurch, Gurchzilla, Gwernol, H2g2bob, Haley92, Hard Raspy Sci, Harryboyles, Hayden120, Hdt83, Hibernian, Hirebrand, Hjal, Hrothberht, Hu12, IanA, Iapetus, Icarus3, IchWeigereMich, IguanaScales, Illusionfx, Informationforall, Invictus the great, Ipatrol, Iroony, J.delanoy, JForget, JNW, JSpung, Jack324, Jackl, Jamarks, Jerimee, Jersey emt, Jezmck, Jfricker, Jimarick, Jmennell, Jmunday1204, Jmundo, JoanAroma, Joe1234, Johnnyio, Johntex, Jojhutton, Jontomkittredge, JorgeGG, Joseph Solis in Australia, Josh Parris, Jossi, Judgesurreal777, Juliancolton, Junglecat, Justin Bailey, Jyngyr, Kalel2007, Kalexandre, Kanags, Kane2742, Kathleenrogers, Kazvorpal, Keichwa, Kelden, Kelsoisme, Kentcamp, Keraunos, Keronjohn0, Kinu, Kmg90, Kmusser, Kmworkman, Knowledge Seeker, Knowledgebycoop, Kugamazog, Kuhlio2008, Kukanotas, Kukini, Kurieeto, KuroiShiroi, Kylepaulmackenzie, LAX, LFaraone, Latka, LegCircus, Legokid, Lightmouse, Lights, Ligulem, Lilac Soul, Longhair, Loren.wilton, Ltbarcly, M C Y 1008, MER-C, MPerel, Mac, Mark Foskey, MarkBuckles, Markie, Marshman, Martin451, Masterdevil101, Matt Yeager, Mattyn0170, Mblumber, McDutchie, Mechanolatry, Megamy, Menchi, Mentifisto, Mh166, Michael Devore, Michael Hardy, Micromaster, Midnightcomm, Miiro, Mike Schwartz, MikeCapone, Minesweeper, Mintleaf, Multixfer, MyNameIsNotBob, Mycroft7, MyrddinEmrys, Nachtsoldat, Nae'blis, Nagy, Navy Blue, Ndunruh, Neilbeach, Neo63, Netsnipe, Nhprman, Ningyo Majo, Nivix, Nlamore, No Guru, Noah Salzman, Noncompliant one, Noneofyourbiznez, Norm mit, Notorious4life, Obradovic Goran, Ohannaiw, Oleg326756, Omar 180, Oneworld25, Oren0, OsamaK, P b1999, Pacificus, PatrikR, Pdculbert, Pearle, Pete Simpson, Peterk92, Petersam, PhilKnight, Philip Trueman, Phoenix2, Phoenixrod, Piano non troppo, Pie7044, Pierre2012, PigFlu Oink, Piledhigheranddeeper, Poslanik, Pradtke, Propaniac, Pumeleon, Pupster21, PureRED, Purgatory Fubar, Pyrospirit, Quatloo, Quintote, QuintusMaximus, RHaworth, Racepacket, RandallJones, RandorXeus, Raso mk, Rdsmith4, Redpoisn01, Reinyday, RenManJoe, Rholton, Rich Farmbrough, Rjwilmsi, Roberta F., Robertb-dc, Rockero, Rootbeer4996, Rrburke, Rwflammang, S3000, Sam Hocevar, Samuella99, Satori Son, Scarebaby, Scepia, Sceptre, SchmuckyTheCat, Scifiintel, Sciurinæ, Scrambler345, Searchme, Senaiboy, Seraphim, Seth Ilys, Shadowlynk, Shawnhath, ShelfSkewed, Sinn, Sistemx, Skew-t, Slakr, Slrubenstein, SmileTodayPublic, Some thing, Songjin, Soosed, Sphivo, Spiffy sperry, SriMesh, StepsTogether, Steven Weston, Steven Zhang, Stewartadcock, Stijl Council, Storm Rider, Stwalkerster, Sun King, Superborsuk, Sverdrup, Svick, Swassociates, TDC, TOttenville8, Tad Lincoln, Tangotango, Tarquin, Techman224, The Illusional Ministry, The JPS, TheRedPenOfDoom, Thespian, Thingg, This, that and the other, Thorpe, Tiddly Tom, Tide rolls, Tillwe, Tohd8BohaithuGh1, Tom harrison, Travelbird, Tregoweth, Tsolosmi, Turangalila, TurboDog1, Tvpm, Tylney1, TzzWzzHzz, Ugen64, Ummit, Unfree, Upinews, User0529, User3, UtherSRG, Vanyagor, Varlaam, Vereinigen, Vince Navarro, Vranak, W Busta, Walter Breitzke, WarrenA, Waterinfo123, Wavelength, Webmediablog, Wes!, Wfpman, WhatamIdoing, Whoisvaibhav, Wik, WikiUserPedia, Wikieditor06, WikipedianMarlith, Will Beback, Willking1979, Wimt, WiscMel, Wizardman, Wmahan, Wolf Of Odin, Wsvlqc, Wtmitchell, Wyzz7, X-20, Xofferson, Xymmax, Yekrats, Yonatan, Z10x, Zachorious, Zen-master, Zensufi, Zhukvita, 968 anonymous edits

Lynn Townsend White, Jr. *Source*: http://en.wikipedia.org/w/index.php?title=Lynn_Townsend_White%2C_Jr. *Contributors*: AdjustShift, Alan Liefting, BD2412, BenBradleyBayhorse, Big iron, CALR, Chanlyn, Cmdrjameson, Cuchullain, Dialectric, Dravecky, Gene Nygaard, HG, Hyarmion, Jheald, JimR, Llywrch, Lockley, Mercury34, Michael A. White, Nathanm mn, Rich Farmbrough, Samw, Savidan, VictoriaHoyle, 20 anonymous edits

Aldo Leopold *Source*: http://en.wikipedia.org/w/index.php?title=Aldo_Leopold *Contributors*: Ahoerstemeier, Alai, Alan Liefting, Alvestrand, Anagramma, Antandrus, Atrian, Bencherlite, Berksee, Bobo192, Bullytr, Calor, CalumH93, Cbarlow, Cdc, Chuunen Baka, Clark89, Cuppysfriend, DMG413, Daniel Collins, Dbsafe, Denisegull, Dhartung, Dj Capricorn, Doug4422, ENeville, Epbr123, Eperotao, Everyking, FJPB, Falcon8765, Feydey, Floridan, ForbiddenWord, FordPrefect42, Gamaliel, Gary Cziko, Greg1016969, Hard Raspy Sci, Hetar, Hraefen, Hrafn, Information Habitat, Iorek85, Ixfd64, JamesMLane, Jamoche, Jeanenawhitney, JimR, Joel Russ, John D. Croft, JohnOwens, JonHarder, JoshuaZ, KAM, KC109, Kevin Forsyth, Kindbudken, Kirkmona, Kirrages, Kukini, Libroman, Lord of the Pit, Maplewooddrive, Message From Xenu, Michaelas10, Mmcannis, Morningstar2651, Nick C, Nlasbo, Nlu, Notheruser, NuclearWarfare, Ottawa4ever, Ouch a cavity, Piano non troppo, Pietro6, Prashanthns, PrenticeCreek, Purgatory Fubar, Queenmomcat, RandallJones, Recognizance, Rror, SDS112, Sandcounty, Severinus, Shoessss, Sift&Winnow, Slyguy, Smeggysmeg, SpeakSinkDive!, Surfeited, TheGrimReaper NS, Viriditas, Whirlingdervish, Yilloslime, Yllosubmarine, Ziel, Zoicon5, ΙωάννηςΚαραμήτρος, 267 anonymous edits

A Sand County Almanac *Source*: http://en.wikipedia.org/w/index.php?title=A_Sand_County_Almanac *Contributors*: Alan Liefting, Apothecia, BrainyBabe, ChristopherM, Doncram, Gidip, GrahamHardy, Gueneverey, Hraefen, Hrafn, Iorek85, JesseStone, Jgrosch, John D. Croft, Jusdafax, Kaolsen, Pegship, Rackettgurl, RandallJones, Steveprutz, Surruk51, The Thing That Should Not Be, Thomp649, Yllosubmarine, 40 anonymous edits

Environmental Values *Source*: http://en.wikipedia.org/w/index.php?title=Environmental_Values *Contributors*: Alan Liefting, Pro bug catcher, Prof C, Wavelength, 3 anonymous edits

Anthropocentrism *Source*: http://en.wikipedia.org/w/index.php?title=Anthropocentrism *Contributors*: 2005-03-07T19:40Z, AC+79 3888, Alan Liefting, Andycjp, BD2412, Bjelkeman, BjörnF, BonsaiViking, Bradjamesbrown, Brooktroutman, Bryan Derksen, Brythain, Debresser, Dfrg.msc, Earthdenizen, Eugenwpg, FT2, Fabiform, Func, George100, Globaleducator, Go for it!, Granitethighs, Gregbard, Grunt, Jaberwocky6669, Jacquerie27, Jheald, Jrockley, KAM, Kaibabsquirrel, Karl Dickman, Kayuki16, Kevlar67, KnightRider, Kooo, LAGoff, La goutte de pluie, Liangbrech, Los3, Mac Davis, Mashford, Mdhowe, Mirrormundo, Mokgand, Motorbyclist, Mwanner, Nectarflowed, Omnipaedista, Owen, Patrickneil, Quebec99, RainbowOfLight, Rd232, Revotfel, Richard001, RickK, Seans33, Sholto Maud, Shousokutsuu, SimonP, Sirius85, Stefeyboy, Sunray, SuperElephant, Tanthalas39, The Transhumanist, Thu, Ucucha, Waggers, Walrasiad, Wgrey, XJamRastafire, 136 anonymous edits

Biocentric individualism *Source*: http://en.wikipedia.org/w/index.php?title=Biocentric_individualism *Contributors*: Anarchia, Common Man, Gregbard, Linas, Pearle, Ringbang, Samxli, Scoty6776, Shanoman, SocratesJedi, Whitepaw, 3 anonymous edits

Biocentrism *Source*: http://en.wikipedia.org/w/index.php?title=Biocentrism *Contributors*: 4ndre4s, Actioncat3, Altarielk, Altenmann, Antarticstargate, Atlant, Barticus88, Debresser, Dogwood123, Egpetersen, Epolk, Fieldday-sunday, Goethean, Hadal, Jeodesic, Jim Yar, Jmdoran, Jordgette, Joyburst, Kaibabsquirrel, Leonard^Bloom, Lesley Fairbairn, Lihaas, Loremaster, Markdandrea, Maxim, Maxmitchell, Mban112, Morning star, Msignor, Muxxa, Nehrams2020, Neutrality, Omnipaedista, Owen, P4k, Pburka, Pengo, Regener, Reviewer4, Ruziklan, Sinneed, Staff3, Stefeyboy, Tobias Schmidbauer, TomLovesCake, Valerius Tygart, Viriditas, WikiWatch31, Wmahan, Zotel, 100 anonymous edits

Conservation (ethic) *Source*: http://en.wikipedia.org/w/index.php?title=Conservation_%28ethic%29 *Contributors*: 0x6D667061, A the 0th, Adriaan, Alan Liefting, All Is One, Andycjp, Anthere, Arthur Rubin, Atulsnischal, AxelBoldt, Beland, BobCMU76, Burlywood, Cactus.man, Cafemusique, Cavrdg, Colonies Chris, Common Man, Cxz111, Cyberbrook, DMacks, Dvknovak, ERcheck, El C, Fashnek, Frymaster, Gabriel Kielland, Gadfium, Gelatinous.Cube, Gidip, Gnomon Kelemen, GrahamN, Grstain, Guanaco, Guettarda, Gunnar Larsson, H2O, Hard Raspy Sci, Husond, Hydro2, IndulgentReader, JRR Trollkien, JerryFriedman, Jiang, Jóna Þórunn, KAM, Kellen`, Koavf, Linmhall, Marcika, Marek69, Mav, Mdd, Memories24, Michael Hardy, Mihoshi, Moguo, Moojoe, Naddy, Nilmerg, Northmeister, Oikoschile, Phlebas, Pigsonthewing, Punchi, Quadell, R Lowry, RK, Renata, Revolución, RexNL, Richard001, Rjwilmsi, SEWilco, Scott Gall, Shawn in Montreal, Shore3, Snoyes, Specs112, Steven Zhang, Stevertigo, Sunray, Syra987, TJRC, TakuyaMurata, Talon Artaine, The Thing That Should Not Be, Thompsma, Vespristiano, Viriditas, Vivio Testarossa, Wapcaplet, Wavelength, Wetman, Wgrey, 101 anonymous edits

Conservation movement *Source*: http://en.wikipedia.org/w/index.php?title=Conservation_movement *Contributors*: AbsolutDan, Alan Liefting, All Is One, Americasroof, Arjuna909, Arthur Rubin, Atulsnischal, Ayanoa, Beetstra, Beland, Bensaccount, Briangogan, Bryan Derksen, Burlywood, CaelumArisen, Camelcast, Carl.bunderson, Citynoise, Ckatz, Common Man, Conservation21, Cwinter, DANIKA27, Dandood333, Disinclination, Durova, Dysmorodrepanis, Ecoconservant, Editore99, Edivorce, El C, Erlendaakre, Fieldday-sunday, Firsfron, Gabriel Kielland, Giraffedata, Goldfishsoldier, Greenopedia, Guanaco, Hadal, Hard Raspy Sci, Hasam, Hu12, Intangible, Irsmart, JRR Trollkien, Jadger, Jatkins, Jeannie kendrick, Jocelyng, John Hill, Johnfos, Johnteslade, KAM, Khardwick, Kingturtle, Kintetsubuffalo, Kurykh, LOL, Lachae115, Laynechavis, Lightmouse, Lokiloki, MER-C, MONGO, Markofu, Martial75, Metta Bubble, Michael Corey, Michael Hardy, Moonriddengirl, Mvp, Neelix, Nom DeGuerre, Npgloma, Oikoschile, PhilipO, Plasticup, Platewq, Radiant!, RainbowOfLight, Regibox, Renata, RevRagnarok, RexNL, Richdagenius, Rishius, Rjensen, Rockfang, SEWilco, Sannse, Shawn in Montreal, Shoeofdeath, Shoessss, Shyamal, Sinneed, SiobhanHansa, Snowmanradio, Steven Walling, Stor stark7, Sunray, Survival123, THEN WHO WAS PHONE?, TastyPoutine, Tearanz, The Thing That Should Not Be, Tide rolls, Track n Field, Tranquilidy, Trilobitealive, Twas Now, Van helsing, Vsmith, Wavelength, Whalewetsuits, Whitejay251, William Avery, YUL89YYZ, Zenohockey, ZooFari, Zuman, 154 anonymous edits

Deep ecology *Source*: http://en.wikipedia.org/w/index.php?title=Deep_ecology *Contributors*: Alan Liefting, AllGloryToTheHypnotoad, Alohasoy, Anarchia, Antandrus, Antarticstargate, ArglebargleIV, Arkuat, Arthur Rubin, Authalic, Ayanoa, Ayecee, B. Bierhoff, B9 hummingbird hovering, Beetstra, Beland, Blueberrypie12, Bobo192, Bookandcoffee, BoundaryRider, Bron Taylor, Bryan Derksen, C mon, Calton, CardinalDan, Catapult, Caveatdumptruck, Cazort, Charles Matthews, Chelseamjackson, Chriswaterguy, Correogsk, CryptoStorm, Cusk1, Dakinijones, Davidkevin, Deeptrivia, Deli nk, Demeter, Dennis Brown, Didier Achermann, Dikjosef, Doc Tropics, Dysprosia, Earthdenizen, Ed Poor, Emmett5, Fishtron, Flamingspinach, FredrikMeyer, Frymaster, Gabbe, Gabriel Kielland, Gadfium, Habiibi, Hearthstone, Heron, Hordaland, Hrafn, Iromeister, James xeno, Jedes, JimR, Jmeppley, Joakim Ziegler, Joel Russ, Joel7687, John Croft, John D. Croft, Joli Rouge, JonasRH, Joyous!, Jrtayloriv, Jtneill, KAM, Kamahi, Kathryn NicDhàna, Kellen`, Kevinccc, Koavf, Lantianer, LaszloWalrus, LindaKaySmith, Losecontrol, MBDowd, Mana Excalibur, Markwalters79, Marnieglickman, Maziotis, Meidosemme, Mejink, Michael Hardy, Mikeybabel01, Mladifilozof, MountainLogic, MrHaiku, MrOllie, MusicScience, Muxxa, Mwanner, NawlinWiki, Nednednerb, NielsenGW, Nightheron9, Nihilo 01, Næss, Olve Utne, Omegatron, Oop, OrionK, Peregrine981, Perklund, Plrk, Plumbago, Poodledog, Portillo, R Lowry,

RJBurkhart3, RK, Reaverdrop, Reinyday, Revotfel, Richard001, Salix alba, Scott D. White, SeanLegassick, Sentience, Sifaka, Simon Fisher, Slady, Smilo Don, Spencerk, Strobilus, Stufam, Subversive element, Sunray, Svick, Swedenborg, Talon Artaine, Tang.josh, Tangerines, Tassedethe, TestPilot, The Ungovernable Force, TheChieftain, TihKov8, TimMony, TomLovesCake, Tomisti, Troypedia, Vanished User 03, Vert et Noir, Viriditas, Warhorus, Wavelength, Wetman, Wgrey, WikiPedant, WolfgangFaber, Yewtree1968, Zweidinge, 183 anonymous edits

Ecocentrism *Source:* http://en.wikipedia.org/w/index.php?title=Ecocentrism *Contributors:* Alan Liefting, Atlant, C mon, Catapult, Dbolton, Esenabre, Garion96, Ginar, Granitethighs, Guettarda, Hologenic, JaGa, Lihaas, Montanastan, Muxxa, Pedro Miguel Camargo da Cunha Rego, Polonium, Rich Farmbrough, Richard001, Salix alba, Sholto Maud, TomLovesCake, Viriditas, WikHead, 16 anonymous edits

Environmental movement *Source:* http://en.wikipedia.org/w/index.php?title=Environmental_movement *Contributors:* A.J.Chesswas, A1DF67, AbsolutDan, AdultSwim, Ahoerstemeier, Alan Liefting, All Is One, Arnavc07, Ayanoa, Batmanand, Beetstra, Bensaccount, BioTube, Boyster, Burlywood, CALR, Catapult, Chwyatt, Ckatz, CommonsDelinker, Crystallina, David Kernow, Dean Morrison, Demi, Devon54, Dr Christopher Heathcote, Dragomiloff, Drernie, Dwalls, EastTN, Ecofreakoftreeland, Ecoman, Ed Poor, Edivorce, Ehouk1, Elagatis, Elfguy, Emperorbma, EntmootsOfTrolls, Erauch, Erick91, Everyking, FISHERAD, Frap, Gabriel Kielland, Gaff, Gaius Cornelius, Ghostfacearchivist, Gilliam, Gregbard, Ground Zero, Grumpyyoungman01, Gubkaandrea, Gurchzilla, Hammbeen, Hard Raspy Sci, Hede2000, Hetar, Hows12, Hu12, Ian Pitchford, Isis, JRR Trollkien, Jeannie kendrick, Jedes, Jheald, JinJ, Jlairdpdx, Joel Russ, John, John Vandenberg, Jorge Stolfi, Juan M. Gonzalez, KAM, Khoikhoi, KimDabelsteinPetersen, Kintetsubuffalo, Kitale, Kpjas, Laurusnobilis, Levineps, Lightmouse, MER-C, Male1979, Marketmou, Martin TB, MartinHarper, Maximus Rex, Michaelorgan, Mimzy1990, Modulatum, Mwanner, Nirvana2013, NorrisMcDonald, Olivier, Pearle, PelleSmith, Pjc51, Platewq, Plinkit, Pollinator, Ragesoss, Raul654, Reinyday, Revolución, Richard001, Rjwellings, Roolearner, Schlüggell, Schmileye, Scott Burley, Sctechlaw, Shain, SidP, Sinneed, Slon02, Snpeck, SpinJ, Stemonitis, Stormscape, Strayson, Syra987, Tabletop, Taka, Tcncv, Tempodivalse, TerriersFan, The Anome, The Ent, The Merciful, Thunderbolt16, Tide rolls, Tillwe, Timiciousknid, TomLovesCake, Tranquilidy, Tree Hugger, Tyson99, UninvitedCompany, Van helsing, Viriditas, Vsmith, WGee, Wavelength, Wetman, Whalewetsuits, WikHead, Wiki alf, Wikiwopbop, Wing Nut, Xeno, Yachtsman1, ZedGreen, 247 anonymous edits

Environmentalism *Source:* http://en.wikipedia.org/w/index.php?title=Environmentalism *Contributors:* 0jam0, 7&6=thirteen, A. B., A94758352, AEMoreira042281, AVRS, Academic Challenger, AdjustShift, Ahoerstemeier, Aitias, Alai, Alan Liefting, Alansohn, Alarob, Albion Tourgee the Younger, Alents, AlexiusHoratius, Alfrado, Alientraveller, All Is One, Alternat2008, Amalas, Amberroom, AndonicO, Andrewlp1991, Andrzej Kmicic, Andycjp, Angela C, Apwilson, Arakunem, Ariobarzan, Arjun01, Art LaPella, Ashenai, Asidemes, Asm76, Athene cunicularia, Atulsnischal, Ayanoa, BMF81, Babecorp, Banaticus, Bathgems, Beetstra, Belladawn, BenFrankliniam, Beno1000, Beta65143, Betacommand, Better environmental evidence, Betterusername, Bigmattl, Bingo234, Biopresto, Bkengland, Bkonrad, Blackgrape99, Bmdavll, Bobby H. Heffley, Bobo192, Bongwarrior, BoogalooDude, Bordello, Bradjamesbrown, Brusegadi, Bugguyak, Burlywood, Bwithh, C mon, C6541, Cactus.man, CallmeLEGEND, Caltas, Capricorn42, Catbar, Centrx, Cgingold, Cgnk, Chasingsol, Chris Capoccia, Circeus, Ckatz, Cleve1212, Climateneutral, Clucz, Cncs wikipedia, Codyfirestorm, Cohesion, Commit charge, CommonsDelinker, CredoFromStart, Cretog8, Cronian, Cunymedia, DHeyward, Dancter, Danjammil, David Latapie, Davidprior, Dawn Bard, DeadEyeArrow, Declutterize, Dekisugi, Dell Adams, DerAnstifter, DerHexer, Dharma8, Dina, Discospinster, Djlayton4, Dmitri Lytov, DocWatson42, Dougofborg, Dporesky, Dr Christopher Heathcote, Dreadstar, Dureo, Dynaflow, Dzied Bulbash, Eassin, Eeekster, Ehouk1, El C, Elflaw, Enviroboy, Envirocorrector, Envirolo232, Erauch, Eric-Wester, Ethicalpublishing, FISHERAD, Fantastic4boy, Farnishk, Findmeee, Fleetflame, FlyingToaster, FrancoGG, Franzose, Frap, Fsotrain09, Fvw, Gabriel Kielland, GabrielVelasquez, Gaff, Galoubet, Geller80, Geologyguy, GeorgeLouis, Gguguvunt, Gilliam, Gman2008, Gogo Dodo, Goldbutt123, Gomm, GreedyCapitalist, Greenbling07, Greenlabelrecords, Grimessm, Ground Zero, Grundle2600, Gugax451, Gveret Tered, Gverstraete, Habj, HaeB, Halon8, Hammbeen, Hard Raspy Sci, Hartz, Hayduke2000, Hellajaxon, HereToHelp, Hfcom, Hi85, HkCaGu, Hows12, Hu12, Ian Pitchford, IchWeigereMich, IIBeDanged, Incantation, Insaneducktaper, Iowawindow, Irishguy, Itubisadiatur, J.delanoy, J04n, JLMadrigal, JNW, JQF, Jacobmalthouse, Jagged 85, Jatkins, Jayantaism, Jedes, Jefffire, Jeffq, Jeffries Scott, Jenniferperryuk, Jheald, Jim Douglas, Jleon, Jmlk17, JodyB, Johnfos, JoshuaZ, Jsysinc, Jtneill, Judicatus, Kafziel, Katana0182, Keilana, Kevinalewis, Killiondude, KimDabelsteinPetersen, Kirksomers123, Kitler005, Kku, Knoepfle32, Kntrabssi, Kreemy, Kurykh, KylaSegala, La goutte de pluie, Lapsed Pacifist, Larklight, Led zec, Levine2112, Lifeissuffering, Lightmouse, Lombarch, Lredman, Luvbug92, MONGO, MaggotA7X, Magician1111, Malibumalice, Mambg, Mandarax, Maralia, Markbuck2000, Marnieglickman, MastCell, Mathnsci, Mavarin, Maxim, Mboverload, Mbralchenko, Mechanically.jam, Mengo, Mennonot, Mercunis, Meredyth, Metalindustrien, Michael Corey, Michael Greiner, Midnightcomm, Mifter, Mike2119, MikeCapone, Mladifilozof, Mortiska, Mr. Billion, Mufka, Munckin, Mwanner, N.Y. Billion, NAHID, Naught101, Neelix, Neil10058, Neparis, New England, Nhprman, Nick carson, Nigosh, Nijelpk, NorrisMcDonald, NuclearWarfare, OMGsplosion, Oda Mari, Oicumayberight, One-world-generation, Onopearls, Organiceverafter, Orion11M87, Osbojos, Osbus, Otolemur crassicaudatus, Owen, Oxymoron83, PKn, Pasky, Patnfm, Paul Erik, Pavel Vozenilek, Pb30, Pbgiv, Pearle, Peehc, Pfly, Piano non troppo, Pirate Thom, Plrk, Plsmith80, Pnastu, Pollinator, Poodledog, Pragmatik, Prof77, Purcelce, R. fiend, RS102704, Rachelnsd, Radiofan14, Radiofreebc, Ragesoss, Raul654, Raz1el, RazielZero, Rbpolsen, Realm of Shadows, Reedy, RepublicanJacobite, Richard001, Rkdudley1, Robert1947, Ronz, RossF18, RoyBlitz, Rpyle731, Rror, Rschmertz, RucasHost, Ryulong, Rzelnik, S.dedalus, SDC, SEWilco, SUL, Salsburg, SasiSasi, Sbritner, Schmidi9, Schmieery, Scott12345678, Selket, Septegram, Shaker john, Sinneed, SiobhanHansa, Skarebo, Skippy-9, Slatham100, Smileagent, Spanner, Spellmaster, Squiddy, Srose, Ssilvers, Staciou, Stone, Styrsman, Sunabozu, Sunray, Swingkid, Syra987, T g7, T0jikist0ni, Taleinfo, Tangerines, Tangotango, Tarret, TastyPoutine, Tb, Tcplanr, Techman224, TerriersFan, Tetrachloromethane, TheEditrix2, Thecompilers, Thegreenpages, Thepathoftruth, They5413, Thing, Threeblindthrice, Tiddly Tom, TimTay, Timdervan, Tobias Schmidbauer, Toddst1, Tom Roberts 55, Tomgally, Toomanysmilies, Tranquilidy, Treyt021, TrogdorPolitiks, Tuvwxyz, Twalters84, Twelvethirteen, Twested, Tyk brkbrkb, Ukexpat, Ulric1313, Urbanimagination, Valentimd, Van helsing, Vdgaitonde, Veritycowper, Vero.Verite, Vishalarora2007, VityUvieu, Vsmith, Wakeyjamie, Wasswascoast, Watch37264, Wavelength, Wghanem, Whalewetsuits, Whirlyman, White&Nerdy21, Wighson, Wikidsoup, WikikIrsc, Wikiwopbop, Wimt, Winterus, Wisco, Wishing girl, Wkdewey, Wnt, Woohookitty, Worthawholebean, Xenbomb, Yankee White, Yilloslime, Yllosubmarine, ZeD unknown, Zepheus, 728 anonymous edits

Environmental skepticism *Source:* http://en.wikipedia.org/w/index.php?title=Environmental_skepticism *Contributors:* A1DF67, Alan Liefting, Arthur Rubin, Can't sleep, clown will eat me, Cmdrjameson, Common Man, Conversion script, Cvazquez09, Dagonet, Dangerous Angel, Dilaudid, Ed Poor, FWBOarticle, Feraljyce, Gctegpipes, Gzuckier, Henry Cassini, Henrygb, Jedes, John Quiggin, Johnfos, Jpatokal, KathrynLybarger, Kurieeto, Lightmouse, MastCell, Nils Simon, Nlarcher, Omphaloscope, Opiaterein, Orangemartin, P Carn, Pearle, Polonium, Punctilius, PxT, QuackGuru, Raygirvan, Rich Farmbrough, Scott Burley, Spiffy sperry, Strayson, TUF-KAT, TheEvilBlueberryCouncil, TotesBoats, UnlimitedAccess, Vespristiano, Vjlenin, Wayland, Weregerbil, William M. Connolley, Yilloslime, 28 anonymous edits

List of environmental philosophers *Source:* http://en.wikipedia.org/w/index.php?title=List_of_environmental_philosophers *Contributors:* 96T, AndonicO, Betacommand, Bjankuloski06en, Corvus cornix, Dcbrigit, Disavian, Eluchil404, Frodeman, Galoubet, Gottliebgottlieb, Gbguffey, Herondance, Jafeluv, Jbaranao, Kevin Forsyth, Knorlock, Koavf, MECU, Michaelpaulnelson, Pokpokgrain, Puddingman11493, Pxma, Riverspell, Rjkirkman, SeamusSweeney, 69 anonymous edits

Van Rensselaer Potter *Source:* http://en.wikipedia.org/w/index.php?title=Van_Rensselaer_Potter *Contributors:* BD2412, BeteNoir, Ceyockey, Correogsk, DS1953, Etacar11, Kevin Forsyth, LexCorp, Open2universe, Rich Farmbrough, RoyFocker, Scolaire, Viriditas, Waacstats

Image Sources, Licenses and Contributors